SAMARITAN ~ CU

BRITISH MEDICAL B   No. 4

BRITIS

British Medical Bulletin is published four times each year, in January, April, July and October.

Subscriptions and single-copy orders should be sent to: Longman Group UK Ltd, Subscriptions Department, Fourth Avenue, Harlow, Essex CM19 5AA.

Subscription rates for 1991 are: £92 (UK) or £105.00/$168.00 (overseas)

Single copies will be available at £29.95 (UK) or £35.00/ $56.00 (overseas)

NEXT ISSUE

Volume 47 No. 1   BREAST DISEASE   *January 1991*

Scientific Editors: A P M FORREST, H J STEWART and T J ANDERSON

Benign breast disease
Classification: The ANDI nomenclature	*L E Hughes*
Cysts & fibroadenomas: Natural history, cancer risk	*J M Dixon*
Management of the painful & nodular breast	*R E Mansel*
Inflammory disease & duct ectasia	*L E Hughes*

Cancer: Diagnosis and management
Curability of breast cancer	*J Haybittle*
The Team approach:	
Diagnosis of palpable lesions	*R W Blamey*
Diagnosis of impalpable lesions	*U Chetty*
Screening for breast cancer	*N E Day*
BSE	*D Mant*
Organisation of a national screening programme	*J Austoker*
Local therapy: options and results	*J R Yarnold*
Adjuvant systemic therapy	*H J Stewart*
Primary systemic therapy	*A P M Forrest*
Psychosexual consideration & effects	*L Fallowfield*

Cancer: The advancing edge
Genesis of cancer	*T J Anderson*
Changing role of the pathologist	*J Sloane*
Practical application of determinants of cell behaviour	*I Ellis*
Genetics and breast cancer	*M Steel/J Clayton*
Oestrogens & breast cancer	
i) Biological considerations	*W R Miller*
ii) Additive hormones	*K McPherson*
Diet	*L J Kinlen*
Prevention including controlateral cancer	*M Baum*

BRITISH
MEDICAL BULLETIN

VOLUME FORTY-SIX
1990

CHURCHILL LIVINGSTONE
EDINBURGH, LONDON, MELBOURNE AND NEW YORK

CHURCHILL LIVINGSTONE
Medical Division of Longman Group UK Limited

Distributed in the United States of America by Churchill
Livingstone Inc., 1560 Broadway, New York, NY 10036, and
by associated companies, branches and representatives
throughout the world.

ISSN 0007-1420
ISBN 0-443-04327-2

CHURCHILL LIVINGSTONE, Medical Division of Longman Group U.K. Limited.
Typeset and printed by H Charlesworth & Co Ltd, Huddersfield

Distributed in the United States of America by Churchill Livingstone Inc., 1560 Broadway,
New York, NY 10036, and by associated companies, branches and representatives through-
out the world. This journal is indexed, abstracted and/or publishd online in the following
media: Current Contents, Scientific Serials Review, Excerpta Medica, USSR Academy of
Science, Biological Abstracts, UMI (Microform), BRS Colleague (full text), Index Medicus,
BIOSIS, NMLUIS, Adonis

© The British Council 1990

ISSN 0007-1420   ISBN 0 443 04327 2

Lipids and Cardiovascular Disease

Scientific Editors: *D J Galton and G R Thompson*

Introduction: Lipids and coronary disease—resolved and
unresolved problems
M F Oliver   865

Apo(a) gene: structure/function relationships and the possible
link with thrombotic atheromatous disease
A Rees, A Bishop & R Morgan   873

Structure and regulation of the LDL-receptor and its gene
A K Soutar & B L Knight   891

Genetic susceptibility to atherosclerosis
J C Chamberlain & D J Galton   917

DNA based diagnostic tests: recombinant DNA and
cardiovascular disease risk factors
A Sidoli, S Galliani & F E Baralle   941

Pathology of atherosclerosis
N Woolf   960

Primary hyperlipidaemia
G R Thompson   986

Secondary hyperlipidaemia
P N Durrington   1005

Treatment of hyperlipidaemia
D R Illingworth   1025

Policies for the prevention of coronary heart disease through
cholesterol-lowering
B M Rifkind   1059

Cholesterol as a risk factor in coronary heart disease
H Tunstall-Pedoe & W C S Smith   1075

Index   1088

1990 Vol. 46 No. 4

Professor D J Galton and Dr J Scott jointly chaired the committee which included Dr G R Thompson that planned this number of the British Medical Bulletin. We are grateful to them for their help, and particularly to Professor D J Galton and Dr G R Thompson who acted as Scientific Editors for the number.

British Medical Bulletin is published by Churchill Livingstone for The British Council, 10 Spring Gardens, London SW1A 2BN

British Medical Bulletin (1990) Vol. 46, No. 4, pp. 865–872
© The British Council 1990

Introduction: Lipids and coronary disease—resolved and unresolved problems

M F Oliver
Wynn Institute for Metabolic Research, London

To my mind, there are more unresolved problems regarding lipids and cardiovascular diseases than those which have been settled. While this may seem disappointing after 40 years of intensive research, the progress which has been made is remarkable and impressive. It is appropriate, therefore, to begin an introduction by outlining the issues which have been resolved and which are more or less internationally agreed. In doing so, I shall deliberately not digress into causes of coronary heart disease (CHD) other than lipids, even though other influences may be as or more important in relation to its pathogenesis.

AREAS OF AGREEMENT

First, an unassailable body of evidence indicates that raised blood cholesterol causes atheroma in the aorta and coronary arteries in many experimental species and, specifically, in man. In general, the higher the concentration the greater the extent of atheromatous involvement. Raised low density lipoproteins (LDL) and reduced high density lipoproteins (HDL) are separately and jointly responsible.

A second fact, very strongly supported by many epidemiological and experimental studies, is that a high dietary intake of saturated fat is a leading cause of high blood cholesterol and of CHD.

A third area of agreement, still very rapidly expanding as is indicated by some of the papers in this issue of the British Medical Bulletin, is that inheritance is a potent determinant of blood cholesterol/lipoprotein concentrations and lipoprotein receptor

0007–1420/90/0046–0865/$10.00

activity: indeed, genetic influences probably determine more than half of the variability in plasma lipoprotein concentrations.

Fourth—but finally for the moment—is the impressive congruity of the benefit of reducing hypercholesterolaemia. All five major primary prevention trials have shown that it is possible to reduce the incidence of CHD in men with initially high plasma concentrations of cholesterol and LDL, and there should no longer be any doubt about the need for aggressive treatment in such men.

OUTSTANDING QUESTIONS

There are so many unresolved issues that it is certain that the basic and clinical scientists of the forthcoming decade will have plenty to study. An outline of these is timely. The zeal of many health educationalists for lowering everyone's cholesterol carries the danger of closing the book on the cholesterol question and with the commercially-based enthusiasm of pharmaceutical companies, questions which the scientific community should address with urgency may be submerged or overlooked. These need to be identified clearly and include the following.

Cholesterol and coronary atheroma

There are many unresolved issues. While the relationship between high plasma LDL and coronary atheroma is undoubted, what is the actual pathogenic role of LDL? What proportion of plasma LDL is taken up by the arterial endothelium via the LDL receptor system? Is unmodified LDL taken up directly by macrophages? What is the relative role of oxidized LDL and how and where does oxidation occur? How do 'scavenger' lipoproteins such as HDL actually remove extracellular cholesterol from the arterial wall? While transfer enzyme systems are well described, little seems to be known about the kinetics or mechanism of excretion of cholesterol from deposits in the arterial wall. Can crystalline cholesterol be mobilized?

A particularly important issue is whether it is possible to cause regression of cholesterol-rich atheromatous deposits in the coronary arteries by lowering LDL cholesterol or raising HDL cholesterol. While there are some studies already suggesting slowing of progression, particularly in patients with heterozygous familial hypercholesterolaemia (FH), results of larger randomized trials are eagerly awaited. Another unresolved problem is how to interpret changes which may consequently occur in the arterial wall. Is it more important, for example, in relation to the function of the coronary arteries in supplying blood to the myocardium to achieve a 50% regression of a 30% occlusive lesion (one which is not normally associated with clinical symptoms) or a 20% regression of a 70% occlusive lesion (usually regarded as clinically important)? It is more likely that the former change would be easier to demonstrate, while the latter may be the really important test of the regression hypothesis. And will successful regression, as distinct from non-progression, necessarily be associated with restoration of the arterial wall to a more or less normal state? Might it not be that the presence of cholesterol over many years, particularly when in the crystalline form, has acted as a sort of scaffold and that its removal might expose that part of the vessel to adverse haemodynamic influences? Successful repair of arterial lesions presumably would be associated with an increase in platelet-fibrin thrombus formation and this might carry an increased risk of small emboli passing downstream into the myocardium.

A related issue, therefore, is to ensure that the pharmaceutical world is not too hasty in interpreting what would appear to be a positive result (if this can be shown) from regression trials as an indication that the clinical manifestations of heart disease will regress to the same extent.

Cholesterol as a risk factor

There are alternative views. One states that elevated plasma cholesterol concentrations, and particularly LDL cholesterol, are the *sine qua non* for the development of CHD and that effectively everyone in developed nations has concentrations too high. This argument is even advanced in explanation of the fact that the commonest cause of death in individuals with the lowest concentration of plasma cholesterol is CHD and it is argued that even these individuals have levels unacceptably high. Those who hold this view believe that everyone should have their plasma cholesterol lowered to levels of 4 mmol/l or less.

The alternative view is that the body requires for cell membrane homeostasis a 'balanced' plasma cholesterol concentration. There is a J relationship between cholesterol and disease with an increase in the incidence of cancer at the lowest end of the normal distribution. It has been proposed, unconvincingly, that most of this relates to incident cancer and is a result of this process. But there are now several studies showing that low cholesterol predicts the development of cancer eight, ten and more years later. Also, there is an inverse relationship between serum cholesterol and haemorrhagic (not thrombotic) stroke. If a large proportion of the cholesterol in the blood of a given individual is genetically determined, cellular homeostasis may also be individually modulated. It is conceivable that the maintenance of normal biological membrane function, in terms of immune resistance, for example, may require, in some individuals, a higher concentration of cholesterol in the plasma than the exponents of the first view would find acceptable.

Other lipids as risk factors

The extent to which raised serum triglycerides and very low density lipoprotein (VLDL) are or are not important in the pathogenesis of CHD is far from resolved. Yet, the lipoprotein moiety specifically related inversely to VLDL—namely HDL—is clearly inversely related to CHD. If hypertriglyceridaemia is important, how is this? Is it a result of an excess of large very low density lipoproteins—if so why? Or is it a function of the fatty acid esterification of VLDL and triglycerides? Is it related to insulin resistance? If raised plasma triglycerides are important, might the adverse effect not be due to a prothrombotic or antifibrinolytic action?

Lipoprotein (a) (Lp(a)) appears to be the most powerful predictor of CHD. It is independent of the common lipoprotein moieties, structurally, genetically, clinically and epidemiologically. Raised Lp(a), or one of its isoforms, may well be an explanation of the not uncommon finding of a young patient with coronary heart disease without any of the orthodox risk factors. Yet we know little about the reasons for its pathogenicity or how to modify this.

Apolipoprotein E polymorphism, particularly the E4 allele, also needs more investigation in order to provide perspective regarding cholesterol absorption from the gut and its relationship to CHD.

Cholesterol lowering and mortality

The five major primary prevention trials and several of the secondary prevention trials leave no doubt that in men reduction of **high** plasma cholesterol levels is associated with reduction of nonfatal myocardial infarction. The degree of reduction of cardiac death is less impressive, possibly because of the numbers recruited in these trials have been insufficient to have the power to show an effect.

An alternative explanation is that the mechansims leading to cardiac death are largely independent of a raised blood cholesterol level. But these trials have also not shown any reduction in non-cardiovascular mortality and we are bound to conclude in 1990 that reducing raised plasma cholesterol does not reduce total mortality. Again, it can be argued that none of the trials have had the power to provide an answer to this question. But we must be careful not to rationalize and the logical step is to wait until new clinical trials have such power. Clinical trials now being conducted using HMG-CoA reductase inhibitors, which will lower cholesterol to a far greater extent than the first generation trials, may show a reduction in total mortality. We will have to wait and see.

Claims have been made that reducing cholesterol does reduce CHD mortality but, interestingly, these come from two studies both using nicotinic acid (the Coronary Drug Project and the Swedish Secondary Prevention Trial). Nicotinic acid has long been known to cardiologists as a vasodilator and, since both studies had a majority of patients with myocardial infarction, it is not impossible that the improved mortality was related to decreased left ventricular work rather than reduction of cholesterol. Again, we need an open mind about this point.

A related issue is whether profound reduction of plasma cholesterol is associated with any adverse effects. The answer to this important question should become apparent with the results of the HMG-CoA reductase trials. Meanwhile, it has to remain on the table because of the possibility of an increase in the incidence of cancer and also of accidents and violence—even if we cannot explain or even be convinced why this should be.

Cholesterol and women

In striking contrast to men, there is no relationship internationally in women between LDL and HDL cholesterol, on the one hand, and CHD mortality, on the other: the correlation coefficient between total cholesterol and CHD mortality in 19 countries with accurate death certification (WHO standards) is $+0.67$ for men and a non-significant $+0.24$ for women. FH may be an exception. There is a very much weaker relationship between plasma cholesterol as a predictor of CHD in women in comparison with men and, effectively, it disappears by the age of fifty. Why is this? One explanation may be that the levels of LDL/HDL needed to produce extensive coronary atheroma do not occur in women until after the menopause and that it may be necessary for such concentrations to be present for 15–20 years to produce CHD. The demonstration of a relationship between lipoproteins and CHD in women would then be difficult because of the confounding and diluting factors of other diseases in women in their seventies. None of the clinical trials have been conducted in women, and it is a false extrapolation to assume that the benefits regarding non-fatal myocardial infarction shown by treatment of middle-aged hypercholesterolaemic men will apply to women. This is an important issue in so far as more than half of the adult population are women. Screening programmes and health education programmes directed vigorously at women may be uneconomic and unjustified.

It does seem fairly clear, however, that oestrogen replacement therapy has reduced the incidence of CHD strikingly. It has yet to be shown whether hormone replacement therapy (opposed oestrogens, as distinct from oestrogens alone) has such a beneficial effect.

Fatty acids and coronary heart disease

Individual fatty acids have individual effects and much more research will be needed to put these in perspective. While it is established that saturated fatty acids in general favour the development of coronary atheroma and CHD, not all have the same actions. Palmitic acid, for example, may have beneficial effects through an antithrombotic action.

Oleic acid not only lowers LDL, but raises HDL and lowers blood pressure. The extent to which linoleic acid has a 'better' effect on lipoproteins than oleic acid has yet to be resolved. Diets rich in saturated fatty acids are by definition deficient in essential fatty acids and more studies will be required to identify the importance of linoleic acid (the principal essential fatty acid) deficiency in communities and individuals with coronary disease. This may become particularly important as emerging evidence is indicating an inverse relationship between CHD and naturally-occurring antioxidants, such as vitamin E and vitamin C. A relative deficiency of essential fatty acids and antioxidants could favour lipid peroxidation and free-radical formation and may even relate to the importance of oxidized LDL.

Some of the polyenes, notably eicosapentaenoic acid (EPA), actually increase LDL cholesterol. Yet EPA, the principal fish fatty acid normally available, probably has an anti-thrombotic action and may favour atheroma regression.

Lipids and thrombosis

One of the most important unresolved areas is the inter-relationship between different lipid moieties and thrombosis/fibrinolysis. While we know that fibrinogen, Factor VII and plasminogen activator inhibitor (PAI) are all increased by raised triglyceride levels, little detail is yet available concerning the nature of this relationship and, specifically, the relationship of individual fatty acids to plasma factors controlling thrombosis. This is true also for the inter-relationship of lipids and platelets. Better control of CHD may come from understanding of this inter-relationship. It is a much neglected area possibly because of the domination of studies of cholesterol-related lipoproteins, and the lack of appropriate methodology to measure intravascular thrombosis.

Age and cholesterol

The relation between cholesterol and CHD decreases in both sexes with advancing age. This may be because most of the cholesterol-related deaths have already occurred but also because other causes of CHD become proportionately more important. Also, diseases other than heart disease become proportionately more common. It has yet to be demonstrated that changing the lifestyle of elderly people or lowering their cholesterol reduces their risk of developing CHD. Perhaps it would be less intrusive for the quality of their lives if they were left alone.

EDUCATION

CHD is not going to disappear as a result of lowering cholesterol and treating abnormal lipoproteins, and those who claim that it will are either naive or ignorant of the pathogenesis of the disease. Therefore, much needs to be done to educate the public correctly and it is the responsibility of those close to the field of lipids and CHD to provide the right balance. The areas of agreement (see start of this introduction) should be clearly and consistently presented. Excessive claims, false expectations and nihilism must all be resisted.

It is also necessary to educate our own profession. There are four risk factors for CHD, not three. They are smoking, raised cholesterol, raised blood pressure and cardiologists. Cardiologists in many countries still need to be convinced of the need for aggressive action to lower high blood cholesterol. One reason for their casual interest is that their duty is to diagnose and treat advanced disease, and many are overwhelmed by the load of clinical problems. Another is scepticism. Others include a legitimate suspicion of public campaigns promoted by self-appointed health educationalists and of the increasing intrusion into clinical judgement by pharmaceutical companies. But physicians should take action against definite hypercholesterolaemia now, be more alert to high risk families and encourage opportunistic screening for raised cholesterol levels.

Many questions remain and the unresolved problems which I have outlined—and others which space does not permit me to address—are considered with scholarship and care by the authors of the chapters in this outstanding issue of the British Medical Bulletin. These reviews provide perspective for the future.

British Medical Bulletin (1990) Vol. 46, No. 4, pp. 873–890

The Apo(a) gene: Structure/function relationships and the possible link with thrombotic atheromatous disease

A Rees
A Bishop
R Morgan
Department Of Medicine, Universty of Wales College of Medicine, Heath Park, Cardiff

The last 25 years have witnessed an exponential increase of interest in the lipoprotein Lp(a). The structure of the gene encoding for its unique apo protein, Apo(a) has been determined resulting in possible structure/function relationships which may explain the close association between elevated levels of Lp(a) and atheromatous vascular disease. These findings may have profound therapeutic implications for the future treatment of the hyperlipidaemias and hypercholesterolaemia in particular.

LIPOPROTEIN(a) [Lp(a)]

The Lp(a) lipoprotein was first described in 1963 by Kare Berg.[1] Unlike other lipoproteins, whose nomenclature is derived from their buoyant density and ultracentrifugation, Berg designated individuals either Lp(a)$^+$ or Lp(a)$^-$ depending on whether precipitin lines were formed on interaction between their sera and antisera from rabbits hyperimmunized with low density lipoprotein (LDL). At the time, Lp(a) was considered to be an autosomal dominant trait but subsequent work, using quantitative immunochemical methods, have shown that Lp(a) represents a quantitative rather than a qualitative genetic marker whose concentration can vary enormously between different individuals.[2,3]

0007–1420/90/0046–0873/$10.00

This variation is considered to be under polygenic control with a major gene effect determining the higher concentrations.[2-4]

Much of the current interest in Lp(a) results from numerous clinical studies which have established a significant correlation between elevated Lp(a) levels and susceptibility to develop coronary artery disease[5-11] and even cerebrovascular disease.[8,12,13] Furthermore, elevated levels of Lp(a) have also been shown to be a strong predictor of saphenous vein graft stenosis after coronary artery bypass surgery.[14] A stepwise increase in mean Lp(a) is found in groups of patients with increasing vein graft stenosis and at Lp(a) levels above 30 mg/dl, 90% of patients demonstrated vein graft stenosis. Whilst the majority of these studies have correlated levels of Lp(a) with coronary artery disease risk in white populations, similar associations have also been reported amongst Japanese subjects.[5,8] However, such results do not necessarily apply to all racial groups. Black subjects in Houston, USA, have average Lp(a) levels almost twice as high as their white counterparts but despite these high levels black males appear to have a lower death rate from coronary heart disease than caucasians.[15] Confirmation that Lp(a) levels are race dependent is found in the observation that Lp(a) concentrations in black natives from the Congo are considerably elevated compared to a matched caucasian population of French extraction.[16] Thus the atherogenic role of Lp(a) in black populations remains to be established.

Despite such exceptions, the racial and ethnic diversity of the caucasian and Japanese populations studied suggests that the association between elevated Lp(a) levels and vascular disease is a ubiquitous phenomenon. In particular, it is evident that if the Lp(a) concentration is above 30 mg/dl, as it is in about 20% of caucasians, the relative risk of coronary atherosclerosis rises two fold. However, when LDL and Lp(a) concentrations are both elevated, the relative risk rise is about five fold.[6] In addition, the association between elevated levels of Lp(a) and coronary heart disease appears stronger in men under 56 years of age[7] and is responsible for much of the familial predisposition observed in individuals.[11]

The mechanisms by which Lp(a) predispose to atherosclerosis are unclear, but recent developments in the molecular and cell biology of Lp(a) has made it the subject of intense research activity resulting in a conceptual link between lipoprotein metabolism, a procoagulant state and the fibrinolytic system.

Structure of the lipoprotein Lp(a)

Lipoprotein(a) is a spherical particle of 250 Angstroms diameter that floats in a density range of 1.05–1.1 g/ml. The lipid composition of lipoprotein(a) closely resembles that of LDL. The protein moiety consists primarily of two distinct proteins, namely the Apo B100 and a unique carbohydrate rich protein, Apo(a). These two proteins are linked by one or more disulphide bonds within the lipoprotein.[17-19] Apo(a) can be separated from Lp(a) by reduction of the disulphide bond(s) linking it to Apo B100 and the residual lipoprotein particle is similar to LDL in many of its physicochemical and immunochemical properties. Furthermore, this particle has similar affinity for the LDL receptor in cultured fibroblasts whilst unreduced Lp(a) has much reduced affinity and capacity for binding and degradation via this receptor.[20] Thus it is concluded that Lp(a) is essentially an LDL particle which is modified by the attachment of Apo(a) to Apo B100 via disulphide bonds. This modification thus confers upon Lp(a) its distinct physico and immunochemical characteristics. The structure of Apo(a) will be discussed later.

The quantitative genetics of Lp(a)

The distribution of plasma levels of Lp(a) [and thus Apo(a)] in caucasian subjects is skewed towards high levels with most values in the lower concentration range (*see* Fig. 1). Statistical genetic studies have advanced several hypotheses to explain the inter-individual variability and intra-individual constancy of Lp(a) levels, most of which conclude that a single autosomal genetic locus is the major determinant of Lp(a) levels.[2-4] Utermann et al. have extensively investigated the inter and intra individual size heterogeneity of the Lp(a) glycoprotein [Apo(a)] in human sera by SDS-polyacrylamide gel electrophoresis and immunoblotting using anti-Lp(a) serum and they have reported an apparent range of molecular weights ranging from 400 000–700 000 Daltons.[21] This size heterogeneity persists despite prior treatment with neuroaminidase, suggesting it is not explained by differences in sialylation of the protein. Several phenotypes have been described based on their relative mobilities compared to Apo B100. These have been categorized into phenotypes F (faster than Apo B100), B (identical to Apo B100), S1, S2, S3 and S4 (slower by differing degrees than Apo B100) and into respective double band phenotypes (see

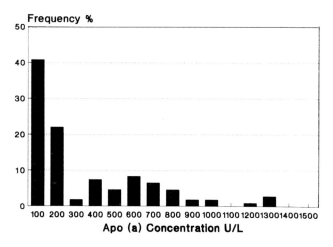

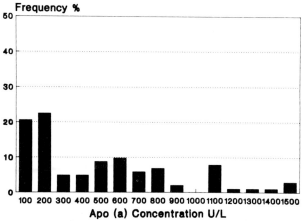

Fig. 1 Frequency distribution of Apo(a) levels in 129 patients with angiographically proven normal coronary arteries (top figure) compared to 102 matched patients with angiographically proven severe coronary artery disease (bottom figure). Difference in distribution is statistically significant, $P < 0.005$.

Fig. 2). Single band phenotypes are considerably more frequent than double band phenotypes with the single band phenotypes S2, S3 and S4 common compared to S1, whilst phenotype B is comparatively rare. 44% of subjects had no detectable Apo(a) protein on immunoblotting due to low or unmeasurable plasma Apo(a) protein concentrations. In view of the individual constancy of the Apo(a) phenotypes together with the fact that no subject had more than two major Apo(a) species in the plasma, Utermann et al. suggest that the Apo(a) polymorphism might be controlled by a

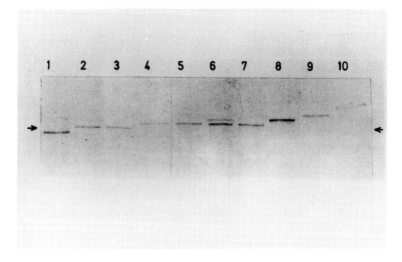

Fig. 2 Immunoblot analysis of Apo(a) phenotypes from individual sera. The following phenotypes are demonstrated, Lanes 1-F; 2,3-B; 4,5 and 7-S1; 6-S1 S2; 8-S2; 9-S3; 10-S4, (*From*: Utermann et al, reference 21).

series of codominant alleles together with an operational null allele. Family studies were in concordance with this hypothesis[22] and it is thus concluded that the Apo(a) phenotypes are controlled by a series of codominant alleles [Lp(a)^F, Lp(a)^B, Lp(a)^{S1}, Lp(a)^{S2}, Lp(a)^{S3}, Lp(a)^{S4}] and the proposed 'null allele' Lp(a)^O at a single major genetic locus. Furthermore, comparison of Lp(a) plasma concentrations in individuals with different phenotypes revealed a highly significant association between phenotype and concentration. Phenotypes B, S1 and S2 associate with high levels of Lp(a) and phenotypes S3 and S4 with low concentrations[23] (see Fig. 3).

Thus there appears to be an inverse correlation between the apparent size of the Apo(a) protein and plasma levels of Lp(a). Such data are highly suggestive that the putative single major locus controlling the Lp(a) glycoprotein phenotypes (i.e. the Apoprotein(a) phenotypes) may also determine Lp(a) levels. The evidence points to this locus being the gene encoding for Apoprotein(a). However, Hasstedt et al.[4] have studied a large Utah pedigree and conclude that genetic variability at a single locus accounts for only

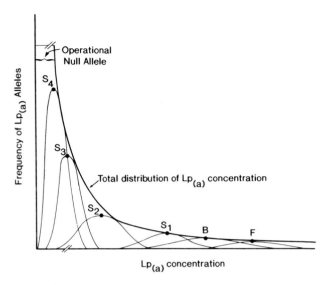

Fig. 3 Relative frequencies of the putative Apo(a) alleles and their association with Lp(a) levels. (*Modified from*: Menzel HJ, Kraft HG, Duba C, Utermann G. Genetic Polymorphism of Lipoprotein(a). In: Steinmetz, Kaffarnik, Schneider. eds. Cholesterol transport systems and their relation to atherosclerosis. Berlin: Springer Verlag, 1989.)

73% of the variability of Lp(a) levels. In addition the population cross sectional studies[21] reveal that within a given phenotype, Lp(a) concentrations vary considerably. Thus it is too simplistic to assume that different Apo(a) allele frequencies are solely responsible for the skewed distribution of Lp(a) concentrations.

Thus, whilst a single genetic locus such as Apo(a) appears to be a major determinant of Lp(a) levels, it is likely that the variability in Lp(a) levels has a polygenic basis in an analogous manner to that observed in other quantitative traits such as triglycerides or cholesterol. The identity of these other genetic loci is currently the subject of much speculation.

STRUCTURE OF APO(a)

Lp(a) closely resembles an LDL particle in lipid composition and in the presence of Apo B100. The distinguishing feature of Lp(a) is the presence of apoprotein(a) bound to Apo B100 by disulphide linkage (see Fig. 4). Apolipoprotein(a) itself is a high molecular weight glycoprotein which exhibits remarkable size heterogeneity with phenotypes ranging in size from 280 000–700 000 Daltons.

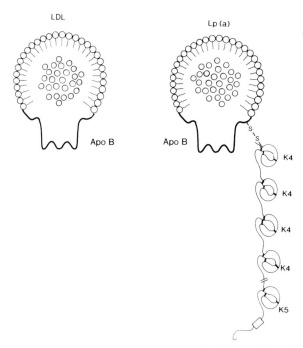

Fig. 4 Schematic representation of the lipoprotein in LDL and Lp(a). Note the Apo(a) peptide linked by a disulphide bridge to apo B in the Lp(a) particle. K4 and K5 donate Kringles 4 and 5 respectively.

The complete primary amino acid sequence of apolipoprotein(a) of unspecified phenotype has been derived by sequencing a cloned human Apo(a) cDNA constructed from a library of hepatic RNAs. A complimentary strategy involves amino acid sequence analysis of the amino terminal end of the intact protein and isolated peptides derived from unlimited proteolysis. Both sequencing strategies result in close agreement and reveal striking sequence homology with plasminogen, the plasmin serine protease zymogen of the fibrinolytic enzyme plasmin.[25,26] Plasminogen is a single chain protein of 791 amino acids containing several distinct structural regions.[27] There is an N-terminal sequence of 76 amino acids followed by 5 distinct domains called Kringles, each exhibiting approximately 40 to 50% homology with each other. These are followed by a trypsin like protease domain.[27] Each Kringle is a cysteine-rich sequence of 80–114 amino acids and includes 6 cysteine residues with disulphide bridges between the 1st and 6th, 2nd and 4th and 3rd and 5th cysteines of each Kringle sequence giving it a characteristic pretzel-like structure reminiscent of a Danish cake called a Kringle. Similar Kringle structures have been observed in prothrombin,[28] tissue plasminogen activator,[29] urokinase[30] and the coagulation factors VII, IX, X, XII and protein C.[31] All of these peptides are considered to be members of a protein super family and are regulatory proteases in both the fibrinolytic and coagulation systems. These regulatory proteases have large non-catalytic segments attached to the amino terminal end of the trypsin-homologue region, a feature that distinguishes them from the simple digestive proteases. In general, the function of these non-catalytic segments, including the Kringle units, is to facilitate binding of the proteases or their zymogens to other macromolecules or receptors, thus mediating and regulating the cascades of fibrinolysis and blood coagulation.[31,32]

Apo(a) gene

The structure of the apoprotein(a) gene and its homology to the plasminogen gene is illustrated in Figure 5. It also reveals distinct structural regions with a hydrophobic signal sequence for secretion, followed by 37 copies of Kringle 4 plasminogen followed by one copy of Kringle 5 and the protease domain all highly conserved with respect to plasminogen. This degree of intragenic homology is practically unique. Of the 37 repeats of the 342 base pair Kringle 4 domain, 24 are identical in nucleotide sequence, an additional 4 differ from the 24 fold repeat in only 3 nucleotides whilst the remaining units differ by 11 to 71 nucleotides. Kringle 3 has a deletion of 24 base pairs whilst Kringle 36 is altered to contain an extra unpaired cysteine and is the likely site of disulphide linkage with Apo B100 in the Lp(a) particle. Although other genes contain internally repeated sequences this number, length and degree of homology is unprecedented and is indicative of relatively recent gene duplication and concomitant expansion and contraction of a locus whose effects can sometimes be seen in populations of the same species. Such an example has been demonstrated in humans where polymorphisms for the number of tandemly arranged haptoglobin related genes have been demonstrated in the haptoglobin gene cluster of blacks.[33] Other examples of unequal but homologous crossing over have been reported in the haemoglobin gene family.

It is thus postulated that the Apo(a) gene originated via gene duplication of the plasminogen gene with deletions of the exons

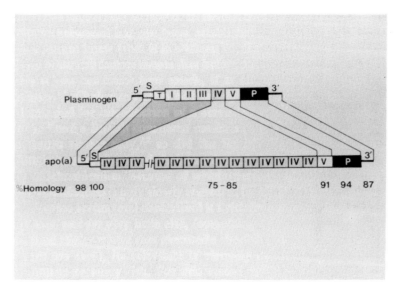

Fig. 5 Schematic representation comparing the sequence of the plasminogen and Apo(a) genes. Each gene is divided into structural domains with the percentage sequence homology shown at the botttom.
Key: 5[1]–untranslated sequence; S–Signal peptide; T–Tail region: I, II, III, IV and V–Kringle domains, I, II, III, IV and V respectively; P–Protease; 3[1]–untranslated sequence (*Modified from*: McLean et al. reference 25).

encoding the NH_2-terminal tail and Kringles 1, 2 and 3. Various untranslated encoding sections of the human Apo(a) gene range from 78% to 100% homology with the plasminogen gene which suggests that an estimated evolution from a common precursor gene occurred approximately 40 million years ago—about the time of divergence of Old and New World monkeys. Whilst DNA hybridization and immunologic data suggest that Apo(a) is restricted to primates, a recent report[34] claims to have identified Apo(a) in the hedgehog. Whether other subprimates possess Apo(a) with highly multiplied Kringle domains and whether these are homologous to their human or rhesus counterparts remains to be established.

Blot hybridization analysis of DNA from a panel of mouse human somatic cell hybrids with an Apo(a) cDNA probe has revealed segregation with chromosome 6 and subsequent in situ hybridization has located both the plasminogen and Apo(a) genes to loci in close proximity on chromosome 6, at 6q 26-27.[35,36]

Further evidence of linkage between the two loci is derived from the studies of linkage in pedigrees between phenotypic variants of the Lp(a) lipoprotein and the plasminogen locus.[37-39] This evidence is suggestive that these two genes probably evolved by the mechanism of gene duplication.

Apo(a) size heterogeneity

The recent evidence suggests that the qualitative variation of Apo(a) observed in human populations is determined by a single major genetic locus which is closely linked to the plasminogen gene on chromosome 6. This locus is most likely to be the Apo(a) gene itself. It is thus postulated that the size heterogeneity observed between different Apo(a) phenotypes represents alternative allelic variants of the Apo(a) gene. The presence of regions of duplicated sequence retaining 100% identity within the gene suggests frequent expansion and contraction of the locus. Thus it is possible that homologous recombination events that add or eliminate Kringle 4 repeats at this locus are probably still occurring and are a likely cause of the size variation of Apo(a) among individuals. Evidence from studies in baboons supports this hypothesis. Baboons, like humans, also possess Apo(a) which exists in distinguishable glycoprotein isoforms and both Apo(a) concentration and phenotypes are similarly genetically determined.[40,41] Hepatic RNA from baboons of known Apo(a) isoforms have been analysed by Northern blot techniques and a variety of Apo(a) transcripts have been detected, ranging in size from 5.2 to 11.2 kilobases.[42] There is a highly significant correlation between transcript size and Apo(a) isoform size, suggesting that the Apo(a) isoforms result from differences in Apo(a) mRNA. However, it is also possible that post-transcriptional modifications also modulate the Apo(a) levels. However, homologous recombination events resulting in varying numbers of tandemly repeated Kringle 4 repeat sequences are the likeliest cause of the intra-individual size variation of the Apo(a) peptide.

Structure–function relationships

Plasminogen is proteolytically inactive until it has been cleaved by tissue plasminogen activator (tPA). However, despite containing a homologous protease domain, Apo(a) is unlikely to possess proteolytic activity as a single nucleotide variation in the Apo(a) gene sequence resulting in a serine-arginine substitution in the amino acid sequence of Apo(a) at the precise cleavage site for tPA precludes its activation to a protease. In addition, Apo(a) from a number of individuals has failed to exhibit any plasmin-like activity, despite treatment with tPA, urokinase and streptokinase.[25] However, the multiple Kringle 4 repeats of Apo(a) suggests, by analogy with other Kringle containing proteins, a function which facilitates or modulates binding to certain receptors, such as the plasminogen receptor or other macromolecules, such as fibrin.

Plasminogen receptors

Plasminogen receptors, originally identified on platelets, have now been identified on a number of peripheral blood cells, including fibroblasts and endothelial cells.[43,44] In particular, endothelial cells play a critical role in thromboregulation via their cell-surface fibrinolytic system (see Fig. 6). Plasminogen receptors are present in extremely high density on endothelial cell surfaces and these receptors promote thrombolysis both by accelerating plasminogen activation by tPA[45] and by protecting plasmin from its inhibitor, alpha-2 antiplasmin.[46] Binding of N-terminal glutamic acid plasminogen (Glu-PLG), the main circulating fibrinolytic zymogen, to the endothelial cell surface receptor results in a 12-fold increase in catalytic efficiency of plasmin generation by tPA, thought to be mediated by conversion of Glu-PLG to its plasmin modified form, N-terminal lysine plasminogen (Lys-PLG).[44] The binding of plasminogen to its receptor is a Kringle dependent phenomenon, raising the possibility that the Kringle-rich Apo(a) peptide may modulate the kinetics of this interaction. A number of studies have addressed this possibility and conclude that Lp(a) competes with plasminogen for its binding site in both a dose-dependent and Kringle-4 dependent manner but has no effect on fluid phase plasmin.[47–49] The physiological relevance of these observations obviously requires further evaluation, but the available evidence suggests that high levels of Lp(a) may alter the delicate balance between profibrinolytic and antifibrinolytic activities in maintaining homeostasis by competing via molecular mimicry with plasminogen for binding sites at the endothelial cell surface. Supporting this hypothesis are the epidemiological data showing that the 20% of the population with levels of Lp(a) greater than 30 mg/dl have a two fold increase in relative risk of developing

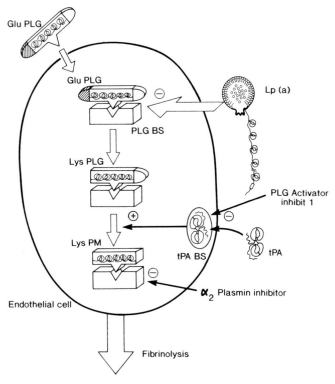

Fig. 6 Schematic representation of the fibrinolytic system on the endothelial cell surface— ⊖ denotes inhibitory effect; ⊕ denotes facilitatory effect.
Key: Glu PLG–N-terminal glutamic acid plasminogen; PLG BS–Plasminogen binding site; Lys PLG–N-terminal lysine plasminogen; PLG Activator inhibit 1– Plasminogen activator inhibitor 1; Lys PM–Lysine plasmin; tPA–Tissue plasminogen activator; tPA BS–Tissue plasminogen activator binding site.

coronary atherosclerosis. If both Lp(a) and LDL-cholesterol levels are elevated, the risk rises to 4 or 5 fold.[6] Using in vitro data from studies on cell cultures, it is estimated that plasma concentrations of Lp(a) of 30 mg/dl would reduce plasminogen binding by 20%, thus encouraging a relatively procoagulant state.[48] Further circumstantial evidence implicating Lp(a) in atherogenesis is the identification of a striking accumulation of Lp(a) immunoreactive material associated with the endothelium and intima of atherosclerotic coronary arteries from autopsy specimens and from the arterial wall of coronary artery bypass patients.[49,50] This immunoreactive material has been proven to be Apo(a) related. Medium-sized arteries with minimal atherosclerotic disease revealed no detectable staining for Lp(a) suggesting that at least in some individuals, Lp(a) accumulates in the vessel wall of atherosclerotic arteries. This may be due to increased binding of Lp(a) to macrophages or arterial wall structures, or it may be secondary to inhibition of fibrinolysis. Whatever the explanation, these tantalizing associations inevitably lead to speculation that Lp(a) provides a bridge between lipid metabolism and thromboregulation at the endothelial surface of the vessel wall, thus providing a potential explanation for the increased thrombotic risks associated with elevated levels of lipoprotein(a).

Metabolism of lipoprotein(a)

Little is known about the synthesis and catabolism of Lp(a). The liver appears to be one obvious source of Lp(a) as the Apo(a) cDNA was constructed from a library of hepatic RNAs.[25] In addition, supporting this contention is the fact that individuals with chronic hepatic disease such as cirrhosis have low levels of Lp(a)[51] and that Apo B100, the other major protein component of Lp(a), is of predominantly hepatic origin.

Lp(a) is the only major lipid risk factor whose concentration appears to be minimally influenced by metabolic, endocrine or anthropometric variables,[52] and the available evidence suggests that although Lp(a) levels are largely determined by a major autosomal genetic locus (presumably the Apo(a) gene itself),[2-4] this may account for only 40% of the variability observed.[24] Whilst this is the strongest effect of a single polymorphic locus on plasma lipid and lipoprotein levels hitherto reported, a number of other gene loci and/or environmental factors may be influential in determining Lp(a) levels. In vivo turnover studies have shown that Lp(a) is catabolised more slowly than LDL and appears not to be derived from either VLDL, LDL or chylomicrons.[53] The contribution of the LDL receptor to the catabolism of Lp(a) in vivo is still controversial. There is disagreement as to whether Lp(a) actually binds to the LDL receptor as the data from in vitro binding studies are conflicting. Several investigators have reported a specific LDL receptor mediated uptake and degradation of native Lp(a) by cultured fibroblasts[20,54,55] although it exhibits reduced affinity to the LDL receptor compared to LDL itself. Reduction and separation of Apo(a) from Lp(a) results in the formation of a remnant particle (Lp(a)⁻) with subsequent restoration of affinity towards the LDL receptor. In addition, the binding, internalisation and uptake of LP(a) into fibroblasts in vitro is independent of the LDL receptor pathway.[56] Cultured human fibroblasts from individuals either heterozygous or homozygous for familial hypercholesterolaemia (FH) showed little difference in affinity for Lp(a) compared to fibroblasts from normal individuals. This is in contrast to the 10 fold difference in affinity observed when LDL particles were studied. In keeping with these in vitro data are the clinical trials reported using different pharmaceutical agents. The HMG CoA-reductase inhibitors competitively inhibit 3-hydroxy-3-methyl-glutaryl (HMG) CoA-reductase, the rate controlling step in cholesterol biosynthesis. This results in a decrease in liver cholesterol synthesis with consequent up regulation of the number of LDL receptors, increased clearance of LDL from plasma and diminution of plasma LDL levels.[57] However, treatment with this class of drug does not reduce Lp(a) levels.[58] Similarly, the cholesterol lowering drug cholestyramine, which also increased liver LDL receptor numbers, has no effect on Lp(a) levels.[59] This suggests that Lp(a) particles, despite being a modified form of LDL, may have a different clearance mechanism from the plasma in vivo. It is possible that the disulphide bonding of the large Apo(a) protein to Apo B of LDL may sterically interfere with LDL receptor binding. In contrast, nicotinic acid and neomycin which are thought to lower LDL levels by decreasing lipoprotein production, have been found to lower Lp(a) levels.[60]

To further confuse matters, patients heterozygous for FH are generally characterized by a 3 fold elevation of Lp(a) levels irrespective of their Apo(a) phenotype and ethnic origin.[61] This is broadly comparable to the 2.5 fold increase in LDL levels typically observed in FH patients.[62] Clearly, therefore, mutations at the LDL receptor gene influence the Lp(a) levels in plasma. Presumably, this locus is one of the putative gene loci contributing to the polygenic determinants of the intra-individual quantitative variability of the Lp(a) trait. However, the precise mechanisms by which this is achieved remain to be established. If Lp(a) is not significantly cleared by the LDL receptor pathway, the increased levels of Lp(a) generally seen in FH patients may be secondary to an increased synthetic rate. Turnover studies are required to address this question. Alternatively, the universal elevation of LDL which concomitantly exists in these patients may inhibit clearance via a number of alternative pathways. Such alternative pathways for clearance of the Lp(a) particle may explain the lack of correlation hitherto observed between LDL levels, Apo B100 levels and Lp(a) levels. This will clearly have important implications for the future treatment of familial hypercholesterolaemia and may reveal an oversimplification in the therapeutic strategies designed for the hyperlipidaemias in general. In the future, management of Lp(a) levels may well become an integral component of a full lipid profile when assessing high risk patients and planning long term therapy.

The understanding of the role of Lp(a) in atherogenesis, the evaluation of the potential synergy between Lp(a) and LDL in promoting ischaemic heart disease and the assessment of the therapeutic benefits of a reduction in Lp(a) levels in specific patient groups should prove a fertile field of study in the near future.

REFERENCES

1 Berg K. A new serum type system in man—the Lp system. Acta Pathol Microbiol Scand 1963; 59: 369–382
2 Harvie NR, Schultz JS. Studies of Lp-lipoprotein as a quantitative genetic trait. Proc Natl Acad Sci USA 1970; 66: 99–103
3 Sing CF, Schultz JS, Shreffler DC. The genetics of the Lp antigen II. A family study and proposed models of genetic control. Ann Hum Genet 1974; 38: 47–56
4 Hasstedt SJ, Wilson DE, Edwards CQ, Cannon WN, Carmelli D, Williams RR. The genetics of quantitative plasma Lp(a): analysis of a large pedigree. Am J Med Genet 1983; 16: 179–188
5 Rhoads GG, Dahlen G, Berg K, Morton NE, Dannenberg AL. Lp(a) lipoprotein as a risk factor for myocardial infarction. J Am Med Assoc 1986; 256: 2540–2544
6 Armstrong VW, Cremer P, Eberle E et al. The association between serum Lp(a) concentrations and angiographically assessed coronary atherosclerosis. Dependence on serum LDL levels. Atherosclerosis 1986; 62: 249–257
7 Dahlen GH, Guyton JR, Attar M, Farmer JA, Kautz JA, Gotto AM. Association of levels of lipoprotein Lp(a), plasma lipids and other lipoproteins with coronary artery disease documented by angiography. Circulation 1986; 74: 758–765
8 Murai A, Miyahara T, Fujimoto N, Matsuda M, Kameyama M. Lp(a) lipoprotein as a risk factor for coronary heart disease and cerebral infarction. Atherosclerosis 1986; 59: 199–204
9 Bishop A, Young TW, Morgan R, Matthews SB, Rees A. Apo(a) levels in patients with and without angiographically defined coronary artery disease. Clin Sci 1989; 77: (suppl No 21) p. 36
10 Berg K, Dahlen G, Frick MH. Lp(a) lipoprotein and pre-β1-lipoprotein in patients with coronary heart disease. Clin Genet 1974; 6: 230–235
11 Durrington PN, Ishola M, Hunt L, Arrol S, Bhatnagar D. Apolipoproteins(a), A1 and B and parental history in men with early onset ischaemic heart disease. Lancet 1988; 2: 1070–1073
12 Jurgens G, Koltringer P. Lipoprotein(a) in ischemic cerebrovascular disease: a new approach to the assessment of risk for stroke. Neurology 1987; 37: 513–515
13 Zenker G, Koltringer P, Bone G, Niederkorn K, Pfeiffer K, Jurgens G. Lipor-

protein(a) as a strong indicator for cerebrovascular disease. Stroke 1986; 17: 942–945

14 Hoff HF, Beck GJ, Skibinski CI et al. Serum Lp(a) level as a predictor of vein graft stenosis after coronary artery bypass surgery in patients. Circulation 1988; 77: 1238–1244

15 Guyton JR, Dahlen GH, Patsch W, Kautz JA, Gotto AM Jr. Relationship of plasma lipoprotein Lp(a) levels to race and to apolipoprotein B. Arteriosclerosis 1985; 5: 265–272

16 Parra H-J, Luyeye I, Bouramoue C, Demarquilly C, Fruchart J-C. Black-White differences in serum Lp(a) lipoprotein levels. Clin Chim Acta 1987; 167: 27–31

17 Utermann G, Weber W. Protein composition of Lp(a) lipoprotein from human plasma. FEBS Lett 1983; 154: 357–361

18 Gaubatz JW, Heideman C, Gotto AM, Morrisett JD, Dahlen GH. Human plasma lipoprotein(a). Structural properties. J Biol Chem 1983; 258: 4582–4589

19 Fless GM, Rolih CA, Scanu AM. Heterogeneity of human plasma lipoprotein(a). J Biol Chem 1984; 259: 11470–11478

20 Armstrong VW, Walli AK, Seidel D. Isolation, characterization, and uptake in human fibroblasts of an apo(a)-free lipoprotein obtained on reduction of lipoprotein(a). J Lipid Res 1985; 26: 1314–1323

21 Utermann G, Kraft HG, Menzel HJ, Hopferwieser T, Seitz C, Genetics of the quantitative Lp(a) lipoprotein trait. I. Relation of Lp(a) glycoprotein phenotypes to Lp(a) lipoprotein concentrations in plasma. Hum Genet 1988; 78: 41–46

22 Utermann G, Duba C, Menzel HJ. Genetics of the quantitative Lp(a) lipoprotein trait. II Inheritance of Lp(a) glycoprotein phenotypes. Hum Genet 1988; 78: 47–50

23 Utermann G, Menzel HJ, Kraft HG, Duba HC, Kemmler HG, Seitz C. Lp(a) glycoprotein: inheritance and relation to Lp(a)-lipoprotein concentrations in plasma. J Clin Invest 1987; 80: 458–465

24 Boerwinkle E, Menzel HJ, Kraft HG, Utermann G. Genetics of the quantitative Lp(a) lipoprotein trait. III. Contribution of Lp(a) glycoprotein phenotypes to normal lipid variation. Hum Genet 1989; 82: 73–78

25 McLean JW, Tomlinson JE, Kuang WJ et al. cDNA sequence of human apolipoprotein(a) is homologous to plasminogen. Nature 1987; 300: 132–137

26 Eaton DL, Fless GM, Kohr WJ et al. Partial amino acid sequence of apolipoprotein(a) shows that it is homologous to plasminogen. Proc Natl Acad Sci USA 1987; 84: 3224–3228

27 Sottrup-Jensen L, Claeys H, Zajdel M, Petersen TE, Magnusson S. The primary structure of human plasminogen: isolation of two lysine-binding fragments and one 'mini'-plasminogen (MW 38 000) by elastase-catalyzed-specific limited proteolysis. In: Davidson JF, Rowan RM, Samama MM, Desnoyers PC, eds. Progress in Chemical Fibrinolysis and Thrombolysis, Volume 3. New York: Raven Press, 1978; pp. 191–209

28 Magnusson S, Petersen TE, Sottrup-Jensen L, Claeys H. Complete primary structure of prothrombin: isolation, structure and reactivity of ten carboxylated glutamic acid residues and regulation of prothrombin activation by thrombin. In: Reich E, Rifkin DB, eds. Proteases and biological control. New York: Cold Spring Harbor, 1975: pp 123–149

29 Pennica D, Holmes WE, Kohr WJ et al. Cloning and expression of human tissue-type plasminogen activator cDNA in E. coli. Nature 1983; 301: 214–221

30 Verde P, Stoppelli MP, Galeffi P, Nocera P, Blasi F. Identification and primary sequence of an unspliced human urokinase poly(A) + RNA. Proc Natl Acad Sci USA 1984; 81: 4727–4731

31 Patthy L. Evolution of the proteases of blood coagulation and fibrinolysis by assembly from modules. Cell 1985; 41: 657–663

32 Patthy L, Trexler M, Vali Z, Banyai L, Varadi A. Kringles: modules specialized for protein binding. Homology of the gelatin-binding region of fibronectin with the Kringle structures of proteases. FEBS Lett 1984; 171: 131–136

33 Maeda N, McEvoy SM, Harris HF, Huisman TH, Smithies O. Polymorphisms in the human haptoglobin gene cluster: chromosomes with multiple haptoglobin-related (Hpr) genes. Proc Natl Acad Sci USA 1986; 83; 7395–7399

34 Laplaud PM, Beaubatie L, Rall SC, Luc G, Saboureau M. Lipoprotein (a) is the major apo B-containing lipoprotein in the plasma of a hibernator, the hedgehog (Erinaceus europaelus) J Lipid Res 1988; 29: 1157–1170

35 Frank SL, Klisak I, Sparkes RS et al. The apolipoprotein (a) gene resides on human chromosome 6q 26–27, in close proximity to the homologous gene for plasminogen. Hum Genet 1988; 79: 352–356

36 Murray JC, Buetow KH, Donovan M et al. Linkage disequilibrium of plasminogen polymorphisms and assignment of the gene to human chromosome 6q 26–6q 27. Am J Hum Genet 1987; 40: 338–350

37 Lindahl G, Gersdorf E, Menzel HJ et al. The gene for the Lp(a)-specific glycoprotein is closely linked to the gene for plasminogen on chromosome 6. Hum Genet 1989; 81: 149–152

38 Weitkamp LR, Guttormsen SA, Schultz JS. Linkage between the loci for the Lp(a) lipoprotein (LP) and plasminogen (PLG). Hum Genet 1988; 79: 80–82

39 Drayna DT, Hegele RA, Hass PE et al. Genetic linkage between lipoprotein(a) phenotype and a DNA polymorphism in the plasminogen gene. Genomics 1988; 3: 230–236

40 Rainwater DL, Manis GS. Immunochemical characterization and quantitation of lipoprotein(a) in baboons. Development of an assay depending on two antigenically distinct proteins. Atherosclerosis 1988; 73: 23–31

41 Rainwater DL, Manis GS, Van de Berg JL. Hereditary and dietary effects on apolipoprotein(a) isoforms and Lp(a) in baboons. J Lipid Res 1989; 30: 549–558

42 Hixson JE, Britten ML, Manis GS, Rainwater DL. Apolipoprotein (a) (Apo(a)) glycoprotein isoforms result from size differences in Apo(a) mRNA in baboons. J Biol Chem 1989; 264: 6013–6016

43 Miles LA, Levin EG, Plescia J, Collen D, Plow EF. Plasminogen receptors, urokinase receptors, and their modulation on human endothelial cells. Blood 1988; 72: 628–635

44 Hajjar KA, Harpel PC, Jaffe EA, Nachman RL. Binding of plasminogen to cultured human endothelial cells. J Biol Chem 1986; 261: 11656–11662

45 Miles LA, Plow EF. Binding and activation of plasminogen on the platelet surface. J Biol Chem 1985; 260: 4303–4311

46 Plow EF, Freaney DE, Plescia J, Miles LA. The plasminogen system and cell surfaces: evidence for plasminogen and urokinase receptors on the same cell type. J Cell Biol 1986; 103: 2411–2420

47 Gonzalez-Gronow M, Edelberg JM, Pizzo SV. Further characterisation of the cellular plasminogen binding site: evidence that plasminogen 2 and lipoprotein(a) compete for the same site. Biochemistry 1989; 28: 2374–2377

48 Miles LA, Fless GM, Levine EG, Scanu AM, Plow EF. A potential basis for the thrombotic risks associated with lipoprotein(a). Nature 1989; 339: 301–303

49 Hajjar KA, Gavish D, Breslow JL, Nachman RL. Lipoprotein (a) modulation of endothelial cell surface fibrinolysis and its potential role in atherosclerosis. Nature 1989; 339: 303–305

50 Rath M, Niendorf A, Reblin T, Dietel M, Krebber HJ, Beisiegel U. Detection and quantification of lipoprotein(a) in the arterial wall of 107 coronary bypass patients. Arteriosclerosis 1989; 9: 579–592

51 Marth E, Cazzolato G, Bittolo Bon G, Avogaro P, Kostner GM. Serum concentrations of Lp(a) and other lipoprotein parameters in heavy alcohol consumers. Ann Nutr Metab 1982; 26: 56–62

52 Sundell IB, Nilsson TK, Hallmans G, Hellsten G, Dahlen GH. Interrelationships between plasma levels of plasminogen activator inhibitor, tissue plasminogen activator, lipoprotein(a), and established risk factors in a North Swedish population. Atherosclerosis 1989; 80: 9–16

53 Krempler F. Kostner GM, Bolzano K, Sandhofer F. Turnover of lipoprotein(a) in man. J Clin Invest 1980; 65: 1483–1490

54 Krempler F, Kostner GM, Roscher A, Haslauer F, Bolzano K, Sandhofer F. Studies on the role of specific cell surface receptors in the removal of lipoprotein(a) in man. J Clin Invest 1983; 71: 1431–1441

55 Havekes L, Vermeer BJ, Brugman T, Emeis J. Binding of Lp(a) to the low density lipoprotein receptor of human fibroblasts. FEBS Lett 1981; 132: 169–173

56 Maartmann-Moe K, Berg K. Lp(a) lipoprotein enters cultured fibroblasts independently of the plasma membrane low density lipoprotein receptor. Clin Genet 1981; 20: 352–362

57 Alberts AW. HMG-CoA Reductase Inhibitors—the development. Atherosclerosis Rev 1988; 18: 123–131

58 Kostner GM, Gavish D, Leopold B, Bolzano K, Weintraub MS, Breslow JL. HMG CoA Reductase Inhibitors lower LDL cholesterol without reducing Lp(a) levels. Circulation 1989; 80: 1313–1319

59 Vessby B, Kostner G, Lithell H, Thomis J. Diverging effects of cholestyramine on apolipoprotein B and lipoprotein Lp(a). A dose-response study of the effects of cholestyramine in hypercholesterolaemia. Atherosclerosis 1982; 44: 61–71

60 Gurakar A, Hoeg JM, Kostner G, Papadopoulos NM, Brewer HB. Levels of lipoprotein Lp(a) decline with neomycin and niacin treatment. Atherosclerosis 1985; 57: 293–301

61 Utermann G, Hoppichler F, Dieplinger H, Seed M, Thompson G, Boerwinkle E. Defects in the low density lipoprotein receptor gene affect lipoprotein(a) levels: multiplicative interaction of two gene loci associated with premature atherosclerosis. Proc Natl Acad Sci USA 1989; 86: 4171–4174

62 Brown MS, Goldstein JL. A receptor-mediated pathway for cholesterol homeostasis. Science 1986; 232: 34–47

British Medical Bulletin (1990) Vol. 46, No. 4, pp. 891–916
© The British Council 1990

Structure and regulation of the LDL-receptor and its gene

A K Soutar
B L Knight
MRC Lipoprotein Team, Hammersmith Hospital, London, UK

The structural features necessary for the efficient
functioning of the LDL receptor are beginning to emerge
from investigation of naturally-occurring and artificially-
produced mutations in the gene. Six of the seven repeated
sequences in the highly-structured NH_2-terminal region are
needed for optimal binding of LDL and some of the
detailed requirements have been elucidated. The
membrane-spanning region is required for insertion of the
protein into the plasma membrane, and the cytoplasmic
region for internalisation and self-association. Many
apparently unrelated mutations affect receptor processing
in the Golgi and the role of the carbohydrate chains
remains obscure. The main means of regulating LDL-
receptor activity is through repression of gene transcription
by sterols. This requires a specific element in the promoter
region and probably involves more than one transcription
factor. Independent effects could be achieved by
modulating the activity of these factors.

The LDL receptor is a glycoprotein present on the surface of most
cells that mediates the uptake and degradation of LDL, the major
cholesterol carrier in plasma. It is responsible for up to 80% of
the clearance of LDL from the circulation, the bulk of which
occurs in the liver, and its activity has a major influence upon the
plasma cholesterol concentration. Since its discovery in 1974 by
Goldstein and Brown, the LDL receptor has been the subject of
intense investigation, which has provided a wealth of information
not only on cholesterol metabolism and heart disease but also on

0007–1420/90/0046–0891/$10.00

the processing and function of cell surface proteins in general. Central to these studies has been the availability of subjects with familial hypercholesterolaemia who have inherited mutant genes that produce defective LDL receptors and so accumulate high concentrations of LDL in their plasma. The general properties of the LDL receptor have been described in previous reviews[1-3] and this article concentrates mainly on the more recent work relating to the identification of the mutations responsible for FH and the way in which they affect the behaviour of the receptor, and to the mechanisms that regulate receptor expression.

RELATIONSHIP BETWEEN STRUCTURE AND FUNCTION

The LDL-receptor is a multifunctional protein

Since the LDL-receptor protein has to carry out several quite distinct functions it should be possible to identify at least some different structural features on the protein that are required for each of these. For example, the protein must contain sufficient information to ensure that it is correctly glycosylated and then transported to and inserted in the cell membrane after its synthesis in the endoplasmic reticulum. It must of course possess specific extracellular high-affinity binding sites for its ligands ApoB and ApoE, and since there is some evidence that suggests that the receptor functions as a dimer or oligomer, there may be additional structural information to facilitate interaction between receptors. Endocytosis of ligand and receptor occurs by movement of receptors into clathrin-coated pits, so the receptor must contain some recognition signal for a protein in the coated pit. After endocytosis the ligand dissociates from the receptor, which then recycles to the cell surface. Since not all cell-surface receptors are recycled, presumably the LDL-receptor has some structural feature required for recycling while ensuring that the ligand is delivered to the appropriate site for degradation. Finally, the activity of the receptor is regulated, in cultured cells at least, by the flux of free cholesterol and the possibility exists that the receptor protein could contain regulatory sites.

Because of its many different functions, it is not surprising that different functional defects can be detected in LDL-receptors in cultured skin fibroblasts from FH patients. Indeed, before any detailed molecular genetic information was available, a classification system for defective mutant LDL-receptors was suggested.[2] This was based on the ability of the receptors to bind LDL and facilitate its internalisation and degradation, and on the characteristic behaviour of newly-synthesized LDL-receptor protein detected by immunoprecipitation from [^{35}S]methionine-labelled cells with specific antibodies. Four classes of mutations were described. Class I mutations are defined as those in which no immunoprecipitable LDL-receptor protein can be detected in cultured skin fibroblasts. This group should now more correctly be subdivided into those mutations which result in no detectable mRNA for the receptor, and those in which mRNA is produced but no protein. These mutations invariably produce a receptor-negative phenotype.

Class II mutations are those in which the newly-synthesized precursor form of the receptor does not undergo the normal process of maturation and translocation to the cell-surface. If the defect in processing is absolute, Class II mutants also produce a receptor-negative phenotype, but if at least some of the mutant protein reaches the cell surface and functions to some extent, then the phenotype will be receptor-deficient.

Class III mutations are those in which an apparently normal protein is synthesized and transported to the cell surface but fails to bind the ligand normally.

Class IV mutations are a group in which the receptor protein is synthesized and appears normally on the cell surface where it is fully capable of binding ligand. However, the mutant receptors fail to cluster in coated pits and are not internalized. Again, depending on the severity of the binding or internalization defect, both Class III and Class IV mutations can result in either a receptor-negative or a receptor-deficient phenotype.

Analysis of the precise changes in protein structure that result in these different functional defects was made possible when cDNA clones for the LDL-receptor were isolated and characterized, permitting prediction of the amino acid sequence of the normal LDL-receptor. This in turn made it possible to identify the change in nucleotide sequence in the LDL-receptor gene in FH patients and so predict the amino acid sequence of the mutant protein.

Domain structure of the LDL-receptor protein

The predicted amino acid sequence of the LDL-receptor protein deduced from the nucleotide sequence of its cDNA, together with information about the arrangement of introns and exons in its gene suggests that the protein comprises five structural regions, or domains, as shown in Figure 1.[4,5] At least two of the domains show extensive homology with different functionally unrelated proteins, which has led to the hypothesis that the LDL-receptor gene has been assembled during evolution by shuffling introns and exons from other older genes, and thus the LDL-receptor protein itself is a mosaic of parts from other proteins. As more and more protein sequence data is predicted by cDNA cloning, it is becoming clear that most large multifunctional proteins share domains in this way.

The first domain of the LDL-receptor encompasses the N-terminal 292 amino acid residues of the protein and comprises seven copies of a cysteine-rich, negatively charged peptide 40 amino acids in length, that shares strong homology, particularly in the spacing of the cysteine residues, with a 40 amino acid sequence found in the human terminal complement component C9.[4] Since it is known that positive charges on ApoB and ApoE interact with the receptor and that reduction of disulphide bonds in the receptor render it unable to bind ligands,[3] it is likely that this domain contains the ligand binding site. Also, the change in electrophoretic mobility that occurs when the protein is reduced suggests that the native receptor contains numerous disulphide bonds that play an important part in maintaining its folded structure.[3]

The second domain comprises approximately 400 amino acids with remarkable homology to the precursor of epidermal growth factor (EGF), including another three cysteine-rich repeats known as growth-factor-like repeats. Even the intron-exon arrangement is identical in the two genes. The third domain comprises the adjacent stretch of 58 amino acids enriched in serine and threonine residues. The majority of the 18 hydroxylated amino acids are

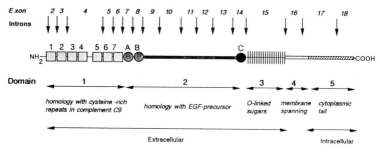

Fig. 1 The domain structure of the LDL-receptor.[4,5]

glycosylated, confirming that this part of the receptor is extracellular. These sugars are added co-translationally and are trimmed and modified as the precursor protein undergoes maturation by passage through the Golgi apparatus to the cell surface. It is this change in carbohydrate residues that results in the apparently anomalous increment in apparent molecular weight that occurs as the protein matures.[2] The fourth domain of the protein is a 22-amino acid stretch of hydrophobic residues bordered by charged residues that is believed to anchor the receptor in the cell surface by spanning the membrane. The fifth and final domain of the LDL-receptor is the COOH-terminal cytoplasmic tail of 50 amino acid residues. The newly-translated protein also contains a cleavable signal sequence that directs the newly synthesized protein to the cell surface.[2]

Thus on the basis of biochemical evidence alone, it is possible to predict what function some of the domains might have, but as will be described in the following sections, more conclusive evidence has come from analysis of the precise nature of the genetic defect in FH patients. In some cases the mutant gene has been transfected into a heterologous cell line to confirm that the observed nucleotide change results in the expected biochemical defect. This is particularly important where the gene of interest is expressed in heterozygous cells where there is one normal and one abnormal gene, or in compound homozygotes. Different mutations have also been introduced into the LDL-receptor gene by mutagenesis in vitro and the constructed genes expressed in transfected cells.

Biosynthesis, maturation and translocation to the cell-surface

Several mutations in the LDL-receptor gene have now been identified that interfere to a greater or lesser extent in the process whereby the newly-synthesized receptor protein is transported to and inserted in the cell-membrane. Examples of some of these are shown in Figure 2.

There are several FH mutants in which no protein can be detected but where a relatively normal amount of LDL-receptor mRNA is found. This implies that the protein is either not synthesized or that it is rapidly degraded. In three of these mutants, the putative protein is a truncated form of the receptor terminating in domain 2. In one case,[6] a 4 kb deletion eliminates exons 13 and

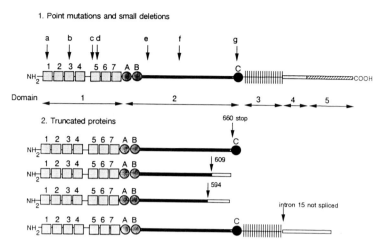

Fig. 2 Mutations in the LDL-receptor gene in FH patients that affect maturation and transport.

All the proteins with small deletions (a[24], b[13], d[13]), single amino acid substitutions (c[11], e[11], f[9], g[12]) or that are truncated in domain 2[6–8] show defective or retarded maturation and transport. The protein with domains 1–3 intact, and domains 4 and 5 deleted matures normally but is not inserted in the cell-membrane and is secreted from the cell.[14–16]

14 and, in another,[7] 5 kb encompassing part of exon 13 until the middle of intron 15 is deleted. In neither of these could a protein be detected. In the Lebanese allele,[8] so named because it is found in a preponderance of the large number of FH patients in that country, a point mutation in exon 14 results in alteration of the codon for amino acid 660 to a stop codon. In this case an immuno-precipitable precursor protein can be detected, but it appears to be trapped in the endoplasmic reticulum (ER) and rapidly degraded. Presumably this is also the fate of the two mutant proteins described above. This suggests that some part of the COOH-terminal part of the protein is essential for transfer of the protein from the ER to the Golgi-apparatus, if only to stabilize its structure. A constructed mutant with a stop codon at the 3′ end of domain 2 is processed and transported normally,[9] although it is secreted from the cell, and a mutant protein from which the whole of domain 2 is deleted is transported normally to the cell-surface.[10] Nonetheless, two other relatively minor mutations in domain 2 result in proteins that are poorly processed. Substitution of valine for glycine at amino acid 544 results in a protein that does not undergo maturation and never reaches the cell surface.[9] Substi-

tution of methionine for valine at 408[11] or of leucine for proline at 664,[12] impedes the transport of newly-synthesized receptor, although receptor protein does eventually appear on the cell surface. Thus it appears that although domain 2 is not essential, if present it must be intact.

Mutations that influence transport are not confined to domain 2 since mutations in domain 1 also result in this phenotype. In the WHHL rabbit, deletion of three amino acids from cysteine-rich repeat 3 and in an FH subject deletion of one amino acid from repeat 5[13] results in mutant LDL-receptors that are modified and transported poorly. At one time it was thought that these two proteins, in which the highly-conserved spacing of cysteine residues in the cysteine-rich repeats is disrupted, might be recognized as foreign because they contain free-SH groups that are normally involved in disulphide bond formation. However, analysis of other mutants has shown that not all transport-defective mutants have disrupted disulphide bonds,[9] and as more and more mutant proteins are identified and characterized, it is surprising how sensitive the process of maturation and transport is to relatively minor changes in structure.

Once the LDL-receptor protein has been transported to the cell surface its insertion into the membrane appears to depend solely on the presence of an intact membrane-spanning domain. Three FH mutations have now been described in which the receptor matures apparently normally but fails to be inserted in the cell membrane and is secreted into the medium. Although the gene deletions that cause the defect are quite different and occur in patients of Finnish,[14] Japanese[15] and American[16] origin, the abnormal protein produced in each case is identical in that it has a normal structure until the end of domain 3, the site for attachment of O-linked sugars, and then 55 random amino acids. The normal cytoplasmic and membrane-spanning domains are absent. However, an intact membrane spanning domain appears to be the only essential sequence for insertion, since mutants in which most or all of domains 1, 2, 3 or 5 have been deleted are inserted normally in the membrane.[10,17–19]

The ligand binding site

The ligand binding site of the LDL-receptor recognizes lipoprotein particles containing either ApoB or ApoE. Particles that contain multiple copies of ApoE bind with higher affinity than those containing either ApoB or a single copy of ApoE, suggesting that particles with multiple ligands either bind to multiple sites on the same receptor or to a single site on each of several receptors.[20] The original observation that the putative ligand binding domain of the LDL receptor predicted from its cDNA sequence contained several identical repeats lent support to the view that each receptor might have more than one binding site. However, mutational analysis showing that relatively minor changes to only a single cysteine-rich repeat can virtually abolish the binding of both types of ligand now suggests that this is unlikely.

Although several mutations in the first domain of the receptor have been identified in the genes of FH patients, in all but one or two of these[10,13,21] it is difficult to determine whether or not binding is defective because maturation and transport is also affected, as discussed above. The most useful information about the structural requirements for ligand binding has come from analysis of constructed mutants. In one study[17] specific mutant cDNAs for the LDL-receptor were expressed in a monkey cell line in which endogenous receptors were fully repressed, and the ability of the transfected cells to bind both LDL and βVLDL was determined as a function of the amount of receptor protein expressed on the cell surface, as assessed by binding of two different monoclonal antibodies. It was reasoned that at least one of the antigenic sites would not be affected by the introduced mutation. Although the details of these experiments are too numerous to describe, taken with observations of naturally-occurring mutations, they permit several conclusions to be drawn, the most important of which is that not all the cysteine-rich repeats in the first domain are of equivalent importance for binding.

First, cysteine-rich repeat 1 is not required for ligand binding, and can be deleted without any deleterious effect on the function of the receptor.[22] However, the anti-receptor monoclonal IgGC7[23] fails to recognize the deleted mutant, suggesting that its antibody antigenic site is dependent upon repeat 1. A naturally-occurring mutation in this first repeat that deletes just two amino acids (aspartate 26, glycine 27), also results in a protein not detected by IgGC7, but in this case receptor function is affected by slow processing and results in an FH phenotype.[24] The results of site-directed mutagenesis largely support these observations, although it is a little surprising that substitution of two cysteines in this repeat with alanine does not appear to affect transport.[17] Unfortunately the effect of the various amino acid substitutions in repeat 1 on binding of

IgGC7 was not documented. This is of interest because the amino acid sequence of cysteine-rich repeat 1 is highly conserved between species, and yet IgGC7 recognizes only the human and bovine receptors and not those of rat, rabbit or hamster.[23] In particular, the glycine and aspartate residues deleted in the FH mutant described above are conserved in all these species.

Secondly, mutations in some of the repeats abolish binding of both ApoB- and ApoE-containing ligands, while in others only binding of LDL is affected and βVLDL is bound normally. This was observed in one of the first naturally-occurring mutations to be identified that specifically affected binding, in which exon 5, which normally codes for the sixth cysteine-rich repeat, was deleted.[21] It is a remarkable feature of this first part of the LDL-receptor gene that the intron-exon junctions and thus the splice sites are all in frame.[5] Thus deletions encompassing exons and parts of their flanking introns always produce proteins in which the whole exon or exons have been deleted without affecting the remaining protein sequence. The mutant with cysteine-rich repeat 6 deleted was able to bind and internalize βVLDL almost normally if binding was expressed relative to the amount of receptor on the cell surface, but was totally unable to bind LDL. It has been speculated that if such differential splicing of exons could occur in vivo, LDL receptor proteins with different ligand specificity could exist in different tissues. From the mutational analysis[17] it was found that single amino acid substitutions in repeats 2, 3, 6 or 7 all had similar effects, in that binding of βVLDL was not impaired, while binding of LDL was impaired or abolished. However, there is an absolute requirement for cysteine-rich repeats 4 and 5 for binding of both βVLDL and LDL. For example substitution of the aspartate residue at 206 in repeat 5 with tyrosine or asparagine virtually abolishes binding of both ligands[17] while only marginally affecting transport. Rather surprisingly, a naturally-occurring mutant has also been described[11] in which substitution of this same aspartate with the relatively similar amino acid residue glutamate has quite marked effects on maturation and transport, and results in a severe FH phenotype.

From these observations it has been suggested[17] that the binding site comprises repeats 2–7 arranged with a two-fold symmetry, as shown in Figure 3. The short stretch of amino acids located between repeats 3 and 4, that is not part of either repeat and is poorly conserved between species, acts as a link between the two halves. However, it is not easy to provide an explanation for the

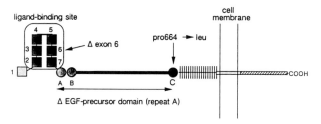

Fig. 3 Possible arrangement of domain 1 cysteine-rich repeats in the ligand-binding site of the LDL-receptor.

The binding site is suggested to comprise repeats 2–7, with a two-fold symmetry on either side of a short stretch of non-conserved amino acids between repeats 4 and 5[17]. Only repeats 4 and 5 are required for binding βVLDL, while repeats 2–7 are essential for normal binding of LDL. Three FH mutations that specifically affect ligand binding are shown.[10,12,21]

observation that ApoE binds readily to receptors containing only the two central repeats, while all six are required for binding of ApoB, when the affinity of the receptor for each individual ligand is similar. Possibly lipoprotein particles containing a single copy of ApoE would not bind to mutants in which all six repeats are not intact.

Although most mutations affecting ligand binding have been localized to domain 1 it has been shown that repeat A in the EGF-precursor-like domain is also required for normal binding.[10] When domain 2 is deleted entirely, the receptor on the cell-surface binds βVLDL but not LDL, but when the receptor is solubilised, fractionated by SDS PAGE and transferred to nitrocellulose, it is able to bind LDL. Thus it is likely that the first cysteine-rich repeat in this domain exerts some constraint on the structure of the binding site in intact cells, but not in the solubilised receptor.

A naturally-occurring point mutation resulting in substitution of proline at 664 with leucine in growth factor repeat C also affects binding,[12] in that it reduces the affinity of binding of both βVLDL and LDL at 4°C and reduces the amount of LDL bound per receptor.[25] It is unlikely that this region, being remote from domain 1, constitutes part of the ligand binding site unless the receptor is folded so that this region is adjacent to domain 1. An alternative explanation is that there is normally some interaction between receptors that is disrupted in this mutant due to alteration in tertiary structure of the protein.

One mutant receptor has been described[26] in which duplication of exons 2–8 has occurred, and the mutant receptor has two complete copies in series of the ligand binding domain plus repeats A and B from the EGF precursor domain. The mutant receptor is apparently translocated normally, despite the fact that its extracellular domain is almost doubled in size, and recycles normally. The increase in apparent molecular weight as the protein precursor matures is the same as that of the normal receptor, suggesting that the sugar modifications responsible for this lie outside the duplicated domains. Since the mutant has two ligand binding domains it could theoretically bind twice as much ligand as normal and twice as much of the monoclonal antibody IgGC7 that recognizes repeat 1. However, since the relationship between the amounts of ligand and antibody bound by the mutant skin fibroblasts is the same as normal, as are the apparent affinities, it is impossible to tell whether both or which, if only one, of the ligand binding sites are accessible.

INTERNALISATION AND RECYCLING

Some of the first mutations in FH subjects to be identified in the genome were those which affected internalisation. Since this process presumably involves some recognition between the receptor and a clathrin-associated protein in the coated pits, it is not surprising that the mutations are located exclusively in the cytoplasmic tail of the receptor protein (Figure 4). In two cases the mutations are single base changes, substituting cysteine for tyrosine at amino acid 807 in one,[27] and introducing a stop codon at position 792 in the other,[18] to give a truncated protein with only two amino acids in the intracellular domain. In the third internalis-

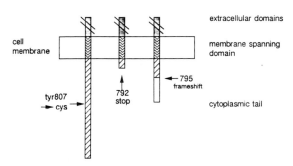

Fig. 4 Mutations in FH patients with internalisation-defective LDL-receptors.
The three mutant LDL-receptor proteins[18,27] are synthesized and transported to the cell-surface where they are inserted into the membrane normally, but they fail to cluster in coated pits and are poorly internalised.

ation-defective allele[18] the protein is normal until residue 795 with an additional eight abnormal residues. A detailed mutational analysis[28] has revealed that the tyrosine residue at 807 is indeed critical. Replacement of this amino acid with an acidic, basic or aliphatic residue abolishes the ability of the receptor to internalise. Tryptophan reduces internalisation by about half, while substitution with phenylalanine results in little loss of activity. The requirements for amino acids on either side of this residue are less stringent, although substitution of proline for the valine at 806 is detrimental. When receptor proteins with truncated cytoplasmic tails of different length were constructed, it was found that deletions after residue 812 were without effect. This is somewhat surprising, since this region is highly conserved between species, and indeed residues 826–839 are identical in receptors from humans, rabbits, hamsters and cows. A recent observation that the receptor may be self-associated on the cell-surface[29] raises the question of to what extent this interaction involves the cytoplasmic tail and is essential for internalisation. Comparison of the susceptibility of various mutants to undergo crosslinking and, by implication, to form oligomers shows that the region close to the membrane is involved. However, there is no evidence that internalisation is dependent on self-association since the mutant receptor truncated at residue 812 is efficiently internalised even though it does not apparently form oligomers.[29]

Once the ligand and receptor complex has been internalised by invagination of coated pits, it is delivered to endosomes where dissociation occurs. The ligand is degraded in secondary lysosomes, while the receptor is recycled to the cell-surface possibly in a vesicle budded off the endosome. Very little is known about the structural features that determine recycling. The only evidence is from a construct in which the whole of domain 2, the EGF-precursor homology domain, has been deleted.[10] This receptor mutant is processed normally and transported to the cell-surface where it binds βVLDL but not LDL. In the absence of ligand the receptor apparently recycles normally, but when ligand has been bound much of the receptor is degraded along with the ligand, and is not recycled. This suggests that dissociation of ligand and receptor in the acidic endosome is impaired in this mutant. What is not certain from these experiments is whether or not there is a specific signal on the receptor for recycling to occur, or whether dissociation of the ligand in the endosome results in automatic recycling. Recently it has been reported[30] that a deletion mutation in a Japanese FH patient which results in the production of a similar, if not identical, mutant LDL-receptor appears to exhibit the same properties.

The LDL-receptor as a glycoprotein

One of the most puzzling aspects of the information that has been gained from analysis of mutants of the LDL-receptor concerns the carbohydrate residues on the protein. The LDL-receptor has both O-linked and N-linked sugars; the O-linked sugars are attached mainly to the third extracellular domain of the receptor adjacent to the membrane-spanning domain, although a few are attached elsewhere, probably to the region of the EGF-precursor homology domain outside the growth factor repeats.[10,31] Consensus sequences for the attachment of N-linked sugars are also found in this domain, although it is not clear which are utilised.[3] When the whole of domain 3 is deleted from the LDL-receptor,[19] the protein functions apparently normally in heterologous cells suggesting that this domain is of no functional significance. However, an FH homozygote has recently been described in whom this same deletion occurs naturally and presumably results in defective function since the FH phenotype is expressed.[32] Also, it is clear from the experiments of Krieger and colleagues, who have isolated by selective mutagenesis a number of mutants of Chinese hamster ovary cells in whom LDL-receptors are absent or defective, that some receptor abnormalities occur in cells in which the only defect is in the trimming and modification of O-linked carbohydrate residues due to deficiency of one of the enzymes concerned.[33] Possibly this domain does not have any functional role in receptor-mediated endocytosis, but may be important in maintenance of receptor protein structure and stability that is not revealed in cultured cells.

REGULATION OF LDL-RECEPTOR ACTIVITY

General

The observation that LDL binding and uptake by cultured human fibroblasts can be increased by removing cholesterol from the cells during incubation with lipoprotein-deficient serum (LPDS) and decreased again on the addition of LDL or free cholesterol was the first demonstration that LDL-receptor activity is under metabolic control.[1] These experiments gave rise to the now familiar concept that LDL-receptor activity is related in some way to the free cholesterol content of the cells. Further studies using antibodies raised against the purified receptor showed that these changes in activity reflected changes in the amount of receptor in the cells. Similar results have subsequently been obtained with a wide variety of cultured cells using a range of extracellular acceptors and donors, and while it is now recognized that different cell types can vary in the extent of their response, regulation by cholesterol still remains the only fully documented mechanism for controlling LDL-receptor activity.

Confirmation that changes in cholesterol balance can affect LDL-receptor activity in vivo has come mainly from experiments with animals. When hepatic demand for cholesterol is increased by feeding adult dogs with bile acid sequestrants[34] or inhibitors of cholesterol synthesis, the plasma cholesterol concentration falls and hepatic membranes show an increase in their ability to bind LDL. Conversely, feeding cholesterol reduces hepatic receptors in these animals.[36] Similar results were obtained with male hamsters, in which the suppression of LDL-receptor-mediated uptake by the liver was shown to be directly related to the cholesterol content of the diet.[37] In man there is a significant inverse correlation between plasma LDL concentration and the LDL-receptor content of the liver[38] and treatment with bile acid sequestrants[39] and inhibitors of cholesterol synthesis[40] increases the rate of receptor-mediated clearance of LDL.

Modulation of receptor activity

Since the LDL receptor is a complex, membrane-bound glycoprotein there is potential for the modulation of binding and uptake through changes in the structure of the receptor itself or of its anchoring membrane. Manipulations that might be expected to alter the properties of cell membranes can have some effect on LDL receptor activity. Thus, in monocyte-derived macrophages the apparent affinity of the receptor for LDL is increased when the cells are pre-incubated with LPDS.[41] However, this is not a physiological manipulation and there is no similar effect on the affinity of the receptor in hepatoma G-2 (Hep G-2) cells.[42] Short-term incubation of fibroblasts with free fatty acids inhibits binding and uptake of LDL,[43] while long-term enrichment of membranes with polyunsaturated fatty acids increases the rate of internalization of LDL-receptors.[44]

It has recently been recognized that some regulatory proteins and membrane receptors act more effectively when associated into dimers or higher oligomers. Some at least, of the receptors on the surface of transformed fibroblasts are close enough to be chemically cross-linked and it has been proposed that LDL receptors can undergo non-covalent association, mediated by self-recognition of a region near the membrane in the cytoplasmic domain.[29] Mutants constructed with cytoplasmic regions truncated close to the membrane are not susceptible to cross-linking but bind LDL normally, indicating that self-association does not influence ligand binding. However, it is possible that the association could be disrupted in the presence of LDL, since apparent cross-linking of receptors with a monoclonal antibody halves the amount of LDL bound.[45] The cytoplasmic region also contains a serine residue at position 833, which can be phosphorylated by a high molecular weight protein kinase containing a heat stable activator that recognizes the LDL receptor.[46] There is evidence that this residue is phosphorylated, at least in the adrenal cortex, but again there is no clear causal relationship between phosphorylation and receptor function.[47]

There is therefore no conclusive evidence to suggest that there is any physiological modulation of LDL-receptor activity once the receptors have been synthesized.

Regulation of receptor synthesis

It has been shown, originally in cultured fibroblasts but now in every cell type studied, that changes in the amount of LDL receptor in the cell are associated with corresponding changes in the rate of receptor synthesis.[2] The mechanism by which receptors are degraded is not yet known, but does not appear to be involved in the regulation of receptor content. The rate of degradation of LDL receptors in skin fibroblasts is decreased in the presence of inhibitors of protein synthesis, suggesting that some short-lived protein can influence or mediates receptor breakdown. However, the rate of degradation remains unaffected by changes in sterol supply that produce a wide range of receptor expression.[48] Thus, changes in LDL-receptor content probably result entirely from alterations in the rate of receptor synthesis. These, in turn, are thought to reflect changes in the rate of transcription of the LDL-receptor gene.[49] The mRNA for the receptor is most abundant in tissues known to express high LDL-receptor activity[50] and its concentration has now been shown to be increased by cholesterol depletion in a wide variety of cultured cells.

LDL-receptor synthesis and mRNA content in cultured cells are decreased by free cholesterol or cholesterol delivered in LDL or other lipoproteins.[3] However, the most potent inhibitors are oxysterols such as 25-hydroxycholesterol or 7-ketocholesterol,[1] raising the possibility that it is not cholesterol itself but an oxysterol either present as an impurity or formed from cholesterol in the cell that is the metabolically active agent. Control of HMG-CoA reductase activity by oxysterols has been extensively investigated and an oxysterol carrier protein identified,[51] but the exact nature of the regulatory sterols and their mechanism of production remains the subject of debate. Although regulation of LDL-receptor activity by oxysterols has received less attention, it is known that ketoconazole, an inhibitor of cytochrome P_{450}-dependent oxidation reactions, increases receptor activity in Hep G-2 cells and prevents lipoproteins from inhibiting transcription of the gene.[52] These observations support the claim for a role for enzymatically produced oxysterols in the regulation of receptor synthesis.

Control of transcription

Selective activation and repression of transcription is mediated by sequence-specific DNA binding proteins that somehow modulate the binding or activity of the transcription complex. The sequences that bind these proteins are usually found in the 5′-flanking region of the gene, upstream of the transcription initiation site. From the study of cells transfected with hybrid genes containing different sections of the 5′-flanking region of the LDL receptor fused to the coding region of a bacterial enzyme, it was found that a fragment that extended 144 base pairs upstream of the initiation site contained all the elements required for expression and sterol regulation of transcription. Systematic scrambling of the sequences in small segments of this region revealed four short sequences that are required for maximum transcription (Figure 5). One is similar to the TATA sequence found in many promoter regions and the other three are imperfect direct 16 base pair repeats that have some homology with the GC-rich region that binds a transcription-activator protein termed Sp1.[53] Two of these repeats (repeats 2 and 3) are immediately adjacent to each other, while the other is some 35 base pairs upstream. DNA footprinting and gel-retardation studies indicate that Sp1 binds to repeats 1 and 3 but not to repeat 2.[54] The region

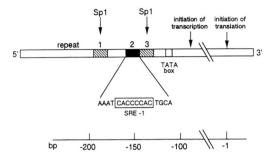

Fig. 5 The promoter region of the LDL-receptor gene.

All the elements required for sterol-regulated transcription of the LDL-receptor gene are located in a region encompassing approx. 150 base pairs upstream from the major initiation site for transcription at the 5′ end of the gene. The region contains three 16-bp repeats and a TATA box.[53] Two of the repeats are binding sites for the transcription factor Spl while eight base pairs in the third repeat confer sensitivity to repression by sterols and have been identified as a sterol response element (SRE-1).[56]

that encompasses repeats 2 and 3 confers sterol regulation upon transcription of a viral gene when inserted into its promoter region.[55] More detailed analysis showed that mutations within repeat 3 prevent transcription of the gene, whereas mutations within repeat 2 abolish the repressive effect of sterols when in tandem with repeat 3.[54] Thus the elements required for sterol-mediated repression are contained within repeat 2. Similar elements adjacent to sites that bind transcription activating factors have been found to be responsible for the effect of sterols on the expression of two other genes involved in cholesterol metabolism, those of HMG-CoA reductase and HMG-CoA synthase. An 8 base pair sequence common to all three elements has been denoted the sterol regulatory element, or SRE-1 (Figure 5).

Studies with hybrid genes transfected into CHO cells have shown that mutations in the SRE-1 element of the HMG-CoA reductase promoter abolish the repressive effect of sterols but have no effect on the high rate of transcription observed in the absence of sterols.[56] Thus in this promoter the SRE-1 element could act simply by binding a protein that directly inhibits transcription or its activation. However, the SRE-1 element in the LDL-receptor gene behaves differently. It is essential for transcription when part of the intact promoter[53] and may in fact exert an independent positive effect in the absence of sterols. If mutant CHO cells that have been selected by their ability to grow in the presence of 25-hydroxycholesterol are transfected with hybrid genes containing the receptor promoter, the rate of transcription of the hybrid gene is high and is not repressed by sterols.[56] Mutations in the SRE-1 region markedly reduce transcription, suggesting that the cells have retained an SRE-1 binding protein that promotes transcription but have lost the ability to block the action of this protein in response to sterols. Despite the different mechanism involved, repression of the HMG-CoA reductase gene by sterols is also lost in the 25-hydroxysterol-resistant cells. Thus the sterol-induced repression of transcription of both genes could be mediated by a single protein whose function is lost in the resistant cells.

The simplest explanation of the results obtained so far is that the SRE-1 element of the LDL-receptor promoter binds a protein factor that permits activation of transcription by Sp1 bound to repeats 1 and 3, and that sterols regulate the ability of a second protein to interfere with the action of this factor. One way in which sterols could exert this effect would be to activate a protein that competes with the positive protein factor for binding to the SRE-1 element. Alternatively the sterol-sensitive protein could modulate the activity of the positive SRE-1 binding factor by, for instance, inducing allosteric changes or combining with it to form an inactive complex. The cDNA sequences of two proteins that could be involved in sterol-mediated repression have recently been elucidated. One is a protein that binds an octanucleotide corresponding to part of the SRE-1 element. Its expression is increased in the presence of sterols but it does not affect the activity of the HMG-CoA reductase promoter in gene constructs.[57] The other is an oxysterol-binding protein which has no obvious DNA binding domain but contains a potential 'leucine-zipper' region that may mediate dimerization to an active complex.[58] A physiological role has yet to be established for either protein.

Inhibition of protein synthesis with cycloheximide has no effect on the initial increase in receptor transcription that occurs when freshly-isolated lymphocytes are incubated with LPDS. Thus the positive activating factors necessary for the initial induction of transcription are already present in the cells. In the longer term protein synthesis is necessary to sustain a high rate of transcription.[59]

Control in the liver

It can be predicted from the response of cultured fibroblasts to extracellular cholesterol and LDL, that the production of LDL-receptors should be repressed in cells in the body. Although this is the case in many tissues, the liver expresses significant numbers of receptors and is responsible for the majority of receptor-mediated clearance of LDL from the circulation. In cells such as fibroblasts, regulation of the expression of LDL receptors is part of a mechanism for maintaing cholesterol homeostasis and providing cholesterol for cell growth and division. The liver has an extra need for cholesterol for the production of bile and lipoproteins and can obtain some of this from the uptake of remnant lipoproteins. Thus the influences upon the metabolically active sterol pool in the liver are likely to be more diverse than those in fibroblasts.

Resistance to the supression of LDL-receptor activity by extracellular LDL and cholesterol is also shown by human hepatocytes[60] and Hep G-2 cells[61] in culture. Further, the weak downregulation by LDL in Hep G-2 cells can be overcome by the addition of HDL, which itself greatly increases receptor expression.[61] In Hep G-2 cells,[62] inhibition of cholesterol esterification greatly enhances the suppression of LDL-receptor and HMG-CoA reductase activities by LDL. These results suggest that the poor response to LDL by liver cells results from diversion of the cholesterol released to cholesteryl esters and external acceptors in preference to the regulatory pools.

Studies with rats and hamsters have shown that hepatic cholesterol synthesis can vary over a 10-fold range in response to changes in cholesterol requirement under conditions when LDL-receptor activity remains constant.[63] Thus in vivo, as in cultured cells, changes in LDL-receptor synthesis are to some extent buffered from major changes in cholesterol supply. This could be achieved either through the relative isolation of a separate regulatory pool or through the action of some sterol-independent mechanism that modifies or over-rides the effect of sterols on receptor expression. There are observations that support both of these possibilities. The reduction in hepatic receptor activity in hamsters fed cholesterol is enhanced by addition of saturated triglycerides to the diet but virtually eliminated by polyunsaturated triglycerides.[37] Similarly unsaturated fat prevents the reduction in hepatic receptor mRNA content of baboons fed cholesterol.[64] There is no obvious change in cholesterol balance and one possible explanation is that the triglycerides cause a redistribution of cholesterol between the various cholesterol pools. On the other hand experiments with human monocyte-macrophages have shown that normal cells express LDL receptors when their rate of cholesterol synthesis is the same as that of cells from FH subjects, suggesting that receptor synthesis may not be regulated exclusively by cholesterol demand.[65]

Hormones and effectors

The action of hormones is one possible mechanism whereby LDL-receptor activity could be regulated independently of sterols. It has long been known that some hormones can influence cellular uptake of LDL and more recent studies have confirmed increases in receptor synthesis and mRNA content. For instance both have been shown to be increased by stimulators of steroid hormone production such as ACTH and chorionic gonadotropin in their target cells[66,67] and by the mitogens PDGF and EGF in cultured fibroblasts.[68] Oestrogen treatment can produce a 10-fold increase in LDL-receptor protein in rat liver[35] but this may not be entirely a direct effect since it gives hardly any stimulation in cultured cells.[69] Hypothyroidism leads to a reduction in hepatic receptor content of rat liver,[70] and thyroxine can stimulate LDL uptake in isolated rat hepatocytes, an effect which is opposed by dexamethasone.[71] LDL uptake by fibroblasts is decreased by epinephrine[72] and increased by insulin.[73] A similar effect of insulin is seen in Hep G-2 cells[42] and rat hepatocytes.[71] Calcium channel blockers[74] and antagonists of calmodulin action[75] increase the LDL-receptor protein and mRNA content of cultured fibroblasts. Phorbol esters, which activate protein kinase C, have little effect on fibroblasts[76] but markedly increase the receptor mRNA content of monocytic leukaemia cells,[77] which also show a reduction of LDL uptake with the prostaglandins PGE_1 and PGE_2.[78] Finally, the hepatic receptor protein content of WHHL rabbits[79] and the receptor mRNA content of human placenta[80] are increased during pregnancy.

Many of these effects can be explained by changes in the requirement of the cells for sterols to support a stimulation of growth. However, some observations are not so readily explained. For instance insulin increases receptor mRNA concentrations in Hep G-2 cells in the presence of maximally suppressive concentrations of LDL.[81] Similarly chorionic gonadotropin stimulates receptor production in granulosa cells in the presence of 25-hydroxycholesterol[67] and calmodulin antagonists further increase receptor synthesis that is already maximally induced by removal of sterols.[76] Thus there is circumstantial evidence for some form of sterol-independent modulation of LDL-receptor synthesis.

In fibroblasts[82], granulosa cells[83] and monocytic leukaemia cells[77] the addition of cycloheximide causes a marked but transient increase in LDL-receptor gene transcription, presumably by preventing the synthesis of rapidly degraded inhibitory proteins. In each case mitogens that would be expected to increase cellular demand for cholesterol produce an additional effect on mRNA production suggesting that these proteins may not be responsive to sterols. At present there is no evidence for the binding of any regulatory factor to the receptor promoter region other than those already described. In granulosa cells the effect of chorionic gondatropin is mimicked by the addition of 8-bromo cyclic AMP.[83] Since the LDL-receptor promoter contains a region apparently similar to cyclic AMP response elements found in other genes, it was proposed that cyclic AMP, or an associated factor, might directly affect receptor transcription. However studies with hybrid genes failed to reveal any such effect.[84]

In Hep G-2[85] and monocytic leukaemia cells[77] dibutyryl cyclic AMP reduces LDL-receptor activity. Thus the reduction in expression by epinephrine and prostaglandins and the increase by insulin could be related, at least in part, to their known abilities to raise and lower the concentration of cyclic AMP in the cell. One of the major effects of cyclic AMP is to stimulate protein kinase activity, which raises the possibility that the action of some regulatory protein is modified by phosphorylation. However, cyclic AMP is also known to stimulate neutral cholesterol esterase activity and so could increase the amount of free sterol available to the regulatory pool. Nevertheless the observation that the concentration of LDL receptor mRNA can also be influenced by calmodulin and phorbol esters, each of which activates a different protein kinase, lends support to the proposal that phosphorylation of transcription factors could play some part in the regulation of LDL-receptor gene expression. Control of receptor expression is clearly more complex than merely the interaction of oxysterols with a binding protein, and the rate of transcription at any time may be determined by the synthesis as well as the modulation of more than one transcription factor or regulatory protein.

REFERENCES

1 Goldstein JL, Brown MS. The low density lipoprotein pathway and its relation to atherosclerosis. Ann Rev Biochem 1977; 46: 897–930
2 Goldstein JL, Brown MS, Anderson RGW, Russell DW, Schneider WJ.

Receptor-mediated endocytosis: concepts emerging from the LDL receptor system. Ann Rev Cell Biol 1985; 1: 1–39
3 Brown MS, Goldstein JL. A receptor-mediated pathway for cholesterol homeostatis. Science 1986; 232: 34–47
4 Yamamoto T, Davis CF, Brown MS et al. The human LDL receptor: a cysteine-rich protein with multiple Alu sequences in its mRNA. Cell 1984; 39: 27–38
5 Südhof TC, Goldstein JL, Brown MS, Russell DW. The LDL receptor gene: a mosaic of exons shared with different proteins. Science 1985; 228: 815–822
6 Horsthemke B, Beisiegel U, Dunning A, Havinga JR, Williamson R, Humphries S. Unequal crossing-over between two alu-repetitive DNA sequences in the low-density-lipoprotein-receptor gene. Eur J Biochem 1987; 164: 77–81
7 Lehrman MA, Russell DW, Goldstein JL, Brown MS. Exon-Alu recombination deletes 5 kilobases from the low density lipoprotein receptor gene, producing a null phenotype in familial hypercholesterolemia. Proc Natl Acad Sci USA 1986; 83: 3679–3683
8 Lehrman MA, Schneider WJ, Brown MS et al. The Lebanese allele at the low density lipoprotein receptor locus. J Biol Chem 1987; 262: 401–410
9 Esser V, Russell DW. Transport-deficient mutations in the low density lipoprotein receptor. J Biol Chem 1988; 263: 13276–13281
10 Davis CG, Goldstein JL, Südhof TC, Anderson RGW, Russell DW, Brown MS. Acid-dependent ligand dissociation and recycling of LDL receptor mediated by growth factor homology region. Nature 1987; 326: 760–765
11 Leitersdorf E, van der Westhuyzen DR, Coetzee GA, Hobbs HA. Two common low density lipoprotein receptor gene mutations cause familial hypercholesterolemia in Afrikaners. J Clin Invest 1989; 84: 954–961
12 Soutar AK, Knight BL, Patel DD. Identification of a point mutation in growth factor repeat C of the low density lipoprotein-receptor gene in a patient with homozygous familial hypercholesterolemia that affects ligand binding and intracellular movement of receptors. Proc Natl Acad Sci USA 1989; 86: 4166–4170
13 Yamamoto T, Bishop RW, Brown MS, Goldstein JL, Russell DW. Deletion in cysteine-rich region of LDL receptor impedes transport to cell surface in WHHL rabbit. Science 1986; 232: 1230–1237
14 Aalto-Setälä K, Helve E, Kovanen PT, Kontula K. Finnish type of low density lipoprotein receptor gene mutation (FH-Helsinki) deletes exons encoding the carboxy-terminal part of the receptor and creates an internalization-defective phenotype. J Clin Invest 1989; 84: 499–505
15 Lehrman MA, Russell DW, Goldstein JL, Brown MS. Alu-Alu recombination deletes splice acceptor sites and produces secreted low density lipoprotein receptor in a subject with familial hypercholesterolemia. J Biol Chem 1987; 262: 3354–3361
16 Lehrman MA, Schneider WJ, Südhof TC, Brown MS, Goldstein JL, Russell DW. Mutation in LDL receptor: Alu-Alu recombination deletes exons encoding transmembrane and cytoplasmic domains. Science 1985; 227: 140–146
17 Esser V, Limbird LE, Brown MS, Goldstein JL, Russell DW. Mutational analysis of the ligand binding domain of the low density lipoprotein receptor. J Biol Chem 1988; 263: 13282–13290
18 Lehrman MA, Goldstein JL, Brown MS, Russell DW, Schneider WJ. Internalization-defective LDL receptor produced by genes with nonsense and frameshift mutations that truncate the cytoplasmic domain. Cell 1985; 41: 735–743
19 Davis CF, Elhammer A, Russell DW et al. Deletion of clustered O-linked carbohydrates does not impair function of low density lipoprotein receptor in transfected fibroblasts. J Biol Chem 1986; 261: 2828–2838
20 Pitas RE, Innerarity TL, Mahley RW. Cell surface receptor binding of phospholipid-protein complexes containing different ratios of receptor-active and -inactive E apoprotein. J Biol Chem 1980; 255: 5454–5460

21 Hobbs HH, Brown MS, Goldstein JL, Russell DW. Deletion of exon encoding cysteine-rich repeat of low density lipoprotein receptor alters its binding specificity in a subject with familial hypercholesterolemia. J Biol Chem 1986; 261: 13114–13120

22 van Driel IR, Goldstein JL, Südhof TC, Brown MS. First cysteine-rich repeat in ligand-binding domain of low density lipoprotein receptor binds Ca^{2+} and monoclonal antibodies, but not lipoproteins. J Biol Chem 1987; 262: 17443–17449

23 Beisiegel U, Schneider WJ, Goldstein JL, Anderson RGW, Brown MS. Monoclonal antibodies to the low density lipoprotein receptor as probes for study of receptor-mediated endocytosis and the genetics of familial hypercholesterolemia. J Biol Chem 1981; 256: 11923–11931

24 Leitersdorf E, Hobbs HH, Fourie AM, Jacobs M, van der Westhuyzen DR, Coetzee GA. Deletion in the first cysteine-rich repeat of low density lipoprotein receptor impairs its transport but not lipoprotein binding in fibroblasts from a subject with familial hypercholesterolemia. Proc Natl Acad Sci USA 1988; 85: 7912–7916

25 Knight BL, Gavigan SJP, Soutar AK, Patel DD. Defective processing and binding of low-density lipoprotein receptors in fibroblasts from a familial hypercholesterolaemic subject. Eur J Biochem 1989; 179: 693–698

26 Lehrman MA, Goldstein JL, Russell DW, Brown MS. Duplication of seven exons in LDL receptor gene caused by Alu-Alu recombination in a subject with familial hypercholesterolemia. Cell 1987; 48: 827–835

27 Davis CG, Lehrman MA, Russell DW, Anderson RGW, Brown MS, Goldstein JL. The J. D. mutation in familial hypercholesterolemia: amino acid substitution in cytoplasmic domain impedes internalization of LDL receptors. Cell 1986; 45: 15–24

28 Davis CG, van Driel IR, Russell DW, Brown MS, Goldstein JL. The low density lipoprotein receptor. Identification of amino acids in cytoplasmic domain required for rapid endocytosis. J Biol Chem 1987; 262: 4075–4082

29 van Driel IR, Davis CF, Goldstein JL, Brown MS. Self-association of the low density lipoprotein receptor mediated by the cytoplasmic domain. J Biol Chem 1987; 262: 16127–16134

30 Miyake Y, Tajima S, Funahashi T, Yamamoto A. Analysis of a recycling-impaired mutant of low density lipoprotein receptor in familial hypercholesterolemia. J Biol Chem 1989; 264: 16584–16590

31 Cummings RD, Kornfeld S, Schneider WJ et al. Biosynthesis of N- and O-linked oligosaccharides of the low density lipoprotein receptor. J Biol Chem 1983; 258: 15261–15273

32 Kajinami K, Mabuchi H, Itoh H et al. New variant of low density lipoprotein receptor gene FH Tonami. Arteriosclerosis 1988; 8: 187–192

33 Kingsley DM, Kozarsky KF, Hobbie L, Krieger M. Reversible defects in O-linked glycosylation and LDL receptor expression in a UDP-Gal/UDP-GalNAc 4-Epimerase deficient mutant. Cell 1986; 44: 749–759

34 Hui DW, Innerarity TL, Mahley RW. Lipoprotein binding to canine hepatic membranes. Metabolically distinct apo-E and apo-B,E receptors. J Biol Chem 1981; 256: 5646–5655

35 Kovanen PT, Brown MS, Goldstein JL. Increased binding of low density lipoprotein to liver membranes from rats treated with 17α-ethinyl estradiol. J Biol Chem 1979; 254: 11367–11373

36 Mahley RW, Innerarity TL. Lipoprotein receptors and cholesterol homeostasis. Biochem Biophys Acta 1983; 737: 197–222

37 Spady DK, Dietschy JM. Interaction of dietary cholesterol and triglycerides in the regulation of hepatic low density lipoprotein transport in the hamster. J Clin Invest 1988; 81: 300–309

38 Soutar AK, Harders-Spengel K, Wade DP, Knight BL. Detection and quantit-

ation of low density lipoprotein (LDL) receptors in human liver by ligand blotting, immunoblotting, and radioimmunoassay. J Biol Chem 1986; 261: 17127–17133

39 Shepherd JS, Packard CJ, Bicker S, Veitch Lawrie TD, Morgan HD. Cholestyramine promotes receptor mediated low density lipoprotein catabolism. N Engl J Med 1980; 302: 1219–1222

40 Malmendier CL, Lontie J-F, Delcroix C, Magot T. Effect of simvastatin on receptor-dependent low density lipoprotein catabolism in normocholesterolemic human volunteers. Atherosclerosis 1989; 80: 101–109

41 Knight, BL, Soutar AK. Low apparent affinity for low-density lipoprotein of receptors expressed by human macrophages maintained with whole serum. Eur J Biochem 1986; 156: 205–210

42 Wade DP, Knight BL, Soutar AK. Hormonal regulation of low-density lipoprotein (LDL) receptor activity in human hepatoma Hep G2 cells. Eur J Biochem 1988; 174: 213–218

43 Bihain BE, Deckelbaum RJ, Yen FT, Gleeson AM, Carpentier YA, Witte LD. Unesterified fatty acids inhibit the binding of low density lipoproteins to the human fibroblast low density lipoprotein receptor. J Biol Chem 1989; 264: 17316–17321

44 Gavigan SJP, Knight BL. Catabolism of low-density lipoprotein by fibroblasts cultured in medium supplemented with saturated or unsaturated free fatty acids. Biochem Biophys Acta 1981; 665: 632–635

45 Gavigan SJP, Patel DD, Soutar AK, Knight BL. An antibody to the low-density lipoprotein (LDL) receptor that partially inhibits the binding of LDL to cultured human fibroblasts. Eur J Biochem 1988; 171: 355–361

46 Kishimoto A, Goldstein JL, Brown MS. Purification of catalytic subunit of low density lipoprotein receptor kinase and identification of heat-stable activator protein. J Biol Chem 1987; 262: 9367–9373

47 Kishimoto A, Brown MS, Slaughter CA, Goldstein JL. Phosphorylation of serine 833 in cytoplasmic domain of low density lipoprotein receptor by a high molecular weight enzyme resembling casein kinase II. J Biol Chem 1987; 262: 1344–1351

48 Casciola LAF, van der Westhuyzen DR, Gevers W, Coetzee GA. Low density lipoprotein receptor degradation is influenced by a mediator protein(s) with a rapid turnover rate, but is unaffected by receptor up- or down-regulation. J. Lipid Res 1988; 29: 1481–1489

49 Russell DW, Yamamoto T, Schneider WJ, Slaughter CJ, Brown MS, Goldstein JL. cDNA cloning of the bovine low density lipoprotein receptor: feedback regulation of a receptor mRNA. Proc Natl Acad Sci USA 1983; 80: 7501–7505

50 Ma PTS, Yamamoto T, Goldstein JL, Brown MS. Increased mRNA for LDL receptor in livers of rabbits treated with 17αethinyl estradiol. Proc Natl Acad Sci USA 1986; 83: 792–796

51 Taylor FR, Saucier SE, Shown EP, Parish EJ, Kandutsch AA. Correlation between oxysterol binding to a cytosolic binding protein and potency in the repression of hydroxymethylglutaryl coenzyme A reductase. J Biol Chem 1984; 259: 12382–12387

52 Takagi K, Alverez JG, Favata MF, Trzoskos JM, Strauss JF. Control of LDL receptor gene promoter activity: ketoconazole inhibits serum lipoprotein but not oxysterol suppression of gene transcription. J Biol Chem 1989; 264: 12352–12357

53 Südhof TC, van der Westhuyzen DR, Goldstein JL, Brown MS, Russell DW. Three direct repeats and a TATA-like sequence are required for regulated expression of the human low density lipoprotein receptor gene. J Biol Chem 1987; 262: 10773–10779

54 Dawson PA, Hofmann SL, van der Westhuyzen DR, Südhof TC, Brown MS, Goldstein JL. Sterol-dependent repression of low density lipoprotein receptor

promoter mediated by 16-base pair sequence adjacent to binding site for transcription factor Spl. J Biol Chem 1988; 263: 3372–3379

55 Südhof TC, Russell DW, Brown MS, Goldstein JL. 42 bp element from LDL receptor gene confers end-product repression by sterols when inserted into viral TK promoter. Cell 1987; 48: 1061–1069

56 Metherall JE, Goldstein JL, Luskey KL, Brown MS. Loss of transcriptional repression of three sterol-regulated genes in mutant hamster cells. J Biol Chem 1989; 264: 15634–15641

57 Rajavashisth TB, Taylor AK, Andalibi A, Svenson KL, Lusis AJ. Identification of a zinc finger protein that binds to the sterol regulatory element. Science 1989; 245: 640–643

58 Dawson PA, Ridgway ND, Slaughter CA, Brown MS, Goldstein JL. cDNA cloning and expression of oxysterol-binding protein, an oligomer with a potential leucine zipper. J Biol Chem 1989; 264: 16798–16803

59 Cuthbert JA, Russell DW, Lipsky PE. Regulation of low density lipoprotein receptor gene expression in human lymphocytes. J Biol Chem 1989; 264: 1298–1304

60 Edge SB, Hoeg JM, Triche T, Schneider PD, Brewer HB. Cultured human hepatocytes. Evidence for metabolism of low density lipoproteins by a pathway independent of the classical low density lipoprotein receptor. J Biol Chem 1986; 261: 3800–3806

61 Havekes LM, Schouten D, de Wit ECM et al. Stimulation of the LDL receptor activity in the human hepatoma cell line Hep G2 by high-density serum fractions. Biochim Biophys Acta 1986; 875: 236–246

62 Havekes LM, de Wit ECM, Princen HMG. Cellular free cholesterol in Hep G2 cells is only partially available for down-regulation of low-density-lipoprotein receptor activity. Biochem J 1987; 247: 739–746

63 Spady SK, Turley SD, Dietschy JM. Rate of low density lipoprotein uptake and cholesterol synthesis are regulated independently in the liver. J Lipid Res 1985; 26: 465–472

64 Fox JC, McGill HC, Carey KD, Getz GS. In vivo regulation of hepatic LDL receptor mRNA in the baboon. J Biol Chem 1987; 262: 7014–7020

65 Patel DD, Pullinger CR, Knight BL. The absolute rate of cholesterol biosynthesis in monocyte-macrophages from normal and familial hypercholesterolaemic subjects. Biochem J 1984; 219: 461–470

66 Faust JR, Goldstein JL, Brown MS. Receptor-mediated uptake of low density lipoprotein and utilisation of its cholesterol for steroid synthesis in cultured mouse adrenal cells. J Biol Chem 1977; 252: 4861–4871

67 Golos TG, August AM, Strauss JF. Expression of low density lipoprotein receptor in cultured human granulosa cells: regulation by human chorionic gonadotropin, cyclic AMP, and sterol. J Lipid Res 1986; 27: 1089–1096

68 Chait A, Ross R, Albers JJ, Bierman EI. Platelet derived growth factor stimulates activity of low density lipoprotein receptors. Proc Natl Acad Sci USA 1980; 77: 4084–4088

69 Semenkovich CF, Ostlund RE. Estrogens induce low-density lipoprotein receptor activity and decrease intracellular cholesterol in human hepatoma cell line Hep G2. Biochemistry 1987; 26: 4987–4992

70 Scarabottolo I., Trezzi E, Roma P, Catapano AL. Experimental hypothyroidism modulates the expression of the low density lipoprotein receptor by the liver. Atherosclerosis 1986; 59: 329–333

71 Salter AM, Fisher SC, Brindley DN. Interactions of triiodothyronine, insulin and dexamethasone on the binding of human LDL to rat hepatocytes in monolayer culture. Atherosclerosis 1988; 71: 77–80

72 Mazière C, Mazière JC, Mora L, Gordette J, Polonovski J. Epinephrine decreases low density lipoprotein processing and lipid synthesis in cultured human fibroblasts. Biochem Biophys Res Commun 1985; 133: 958–963

73 Chait A, Bierman EL, Albers JJ. Regulatory role of insulin in the degradation of low density lipoprotein by cultured human skin fibroblasts. Biochim Biophys Acta 1978; 529: 292–299
74 Filipovic I, Buddecke E. Calcium channel blockers stimulate LDL receptor synthesis in human skin fibroblasts. Biochem Biophys Res Commun 1986; 136: 845–850
75 Eckardt H, Filipovic I, Hasilik A, Buddecke E. Calmodulin antagonists increase the amount of mRNA for the low-density-lipoprotein receptor in skin fibroblasts. Biochem J 1988; 252: 889–892
76 Filipovic I, Buddecke E. Calmodulin antagonists stimulate LDL receptor synthesis in human skin fibroblasts. Biochim Biophys Acta 1986; 876: 124–132
77 Auwerx JH, Chait A, Deeb SS. Regulation of the low density lipoprotein receptor and hydroxymethylglutaryl coenzyme A reductase genes by protein kinase C and a putative negative regulatory protein. Proc Natl Acad Sci USA 1989; 86: 1133–1137
78 Krone W, Klass A, Nägele H, Behnke B, Greten H. Effects of prostaglandins on LDL receptor activity and cholesterol synthesis in freshly isolated human mononuclear leukocytes. J Lipid Res 1988; 29: 1663–1669
79 Shiomi M, Ito T, Watanabe Y. Increase in hepatic low-density lipoprotein receptor activity during pregnancy in Watanabe heritable hyperlipidemic rabbits; an animal model for familial hypercholesterolemia. Biochim Biophys Acta 1987; 917: 92–100
80 Furuhashi M, Seo H, Mizutani S, Narita O, Tomoda Y, Matsui N. Expression of low density lipoprotein receptor gene in human placenta during pregnancy. Mol Endocrinol 1989; 3: 1252–1256
81 Wade DP, Knight BL, Soutar AK. Regulation of low-density-lipoprotein-receptor mRNA by insulin in human hepatoma Hep G2 cells. Eur J Biochem 1989; 181: 727–731
82 Mazzone T, Basheeruddin K, Duncan H. Inhibitors of translation induce low density lipoprotein receptor gene expression in human skin fibroblasts. J Biol Chem 1989; 264: 15529–15534
83 Golos TG, Strauss JF, Miller WL. Regulation of low density lipoprotein receptor and cytochrome P-450scc mRNA levels in human granulosa cells. J Steroid Biochem 1987; 27: 767–773
84 Takagi K, Hoffman EK, Strauss JF. The upstream promoter of the human LDL receptor gene does not contain a cyclic AMP response element. Biochem Biophys Res Commun 1988; 152: 143–148
85 Mazière C, Mazière JC, Salmon S et al. Cyclic AMP decreases LDL catabolism and cholesterol synthesis in the human hepatoma line Hep G2. Biochem Biophys Res Commun 1988; 156: 424–431

British Medical Bulletin (1990) Vol. 46, No. 4, pp. 917–940
© The British Council 1990

Genetic susceptibility to atherosclerosis

J C Chamberlain
D J Galton
Department of Human Genetics and Metabolism, Medical Professorial Unit, St. Bartholomew's Hospital, London UK

Genetic factors are implicated in atherogenesis by family and twin studies of coronary artery disease, interacting with the environment to produce the phenotypic disease. Restriction fragment length polymorphisms provide useful linkage markers with which to study the genetics of this disease, the effectiveness of marker loci being characterized in terms of their polymorphism information content.

Other forms of nucleotide variation, including variable number of tandem repeats can also provide linkage markers for aetiological loci and can be detected by the use of polymerase chain amplification followed by sequencing, denaturing gel electrophoresis, base pair specific chemical cleavage or the use of oligonucleotide probes. Linkage markers may be used either in population association or familial studies. Candidate genes may be studied or complete genomic mapping attempted.

A review of potential candidate genes for atherosclerosis is presented.

Atherosclerosis is a major cause of death in Western Society and its pathology is as complex as it is important. There is no one agent responsible for all atherogenesis, it is mostly a multifactorial disease. It can be the end product of many influences, both environmental and genetic, but prediction is ineffective due to difficulties in the relative weighting of these factors for any individual or population.

Progress is likely to be made by limiting early environmental risk factors in individuals with a specific genetic predisposition. This review describes genetic markers of susceptibility to athero-

0007–1420/90/0046–0917/$10.00

sclerosis and shows how they aid our understanding of the underlying pathology.

The aggregation of premature coronary artery disease within families is well established.[1] Slack and Evans[2] analysed first degree relatives of 121 men and 96 women with coronary artery disease and showed the increased risk of death from coronary artery disease was five and seven fold greater than in matched controls for males and females respectively. Since then many excellent studies have confirmed this trend.[3,4] Further evidence for the implication of a genetic predisposition to atherosclerosis comes from twin studies.[5] Berg found concordance rates for angina pectoris or myocardial infarction in monozygotic twins to be 0.65 as compared with 0.25 for dizygotic pairs. If twins with premature coronary artery disease appearing before age 60 were alone considered these figures were 0.83 and 0.22 respectively.[6]

There is then, more than sufficient evidence to implicate genetic factors in the development of premature coronary atherosclerosis and cardiovascular disease but the evidence for extracoronary disease is nowhere near as convincing, though evidence for the shared molecular pathogenesis of the two diseases is supported by the findings of the Framingham study[7] and the Prospective Basle Study[8] linking the occurrence of coronary artery disease and claudication in their respective populations.

In considering the inheritance of atherosclerosis it is useful to bear in mind its multifactorial nature. This may be expressed in terms of a Venn model (Fig. 1). The sets define subgroups of the general population, **A** defines those members of the population exposed to a highly atherogenic environment. **B** represents a subset composed of individuals exposed to a higher than background atherogenic risk due to a disorder known to predispose to atherogenesis such as hypertension or diabetes mellitus. The remaining subgroup **C** represents those possessing genetic variants conferring a predisposition to develop atherosclerosis. When this genetic liability coincides with other risk factors atheroma will develop.

Such a model can help explain some of the more puzzling features of the genetics of atherosclerosis. Firstly the relatively high frequency of the disease in West European populations may be related to the fact that genes conferring susceptibility to atheroma would be at no selective disadvantage under favourable environmental conditions and would thus be expected to be widespread in healthy subgroups. Secondly in the face of such a complex model the failure of classical genetic epidemiology to elucidate

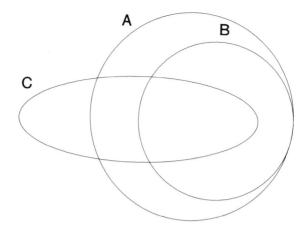

Fig. 1. Venn diagram demonstrating factors that may interact to produce athero-sclerosis. Subset A: members exposed to atherogenic environment; Subset B: individuals with disorder predisposing to atherosclerosis. Subset C: individuals with genetic predisposition.

recognizable patterns of disease inheritance becomes easily explained. Only by the use of the 'new genetics' of recombinant DNA technology does it become possible to study the genes suspected to be involved in the disease process (the so called 'candidate genes') without recourse to a phenotypic intermediate before the end-stage disease.

EXPERIMENTAL APPROACHES

Restriction fragment length polymorphisms

The demonstration of such polymorphism depends on the production of recombinant DNA probes for the area of the genome under study and involves the use of naturally occurring type II restriction endonucleases, which will cleave DNA only in relation to specific nucleotide sequences. The action of such enzymes on extracted human genomic DNA results in a mixture of variously sized fragments of DNA which may be fractionated according to size by agarose gel electrophoresis. These sized fragments may then be transferred onto a hybridization filter[9] and the relevant fragments identified by hybridization with the labelled DNA probe mentioned previously.

This technique can reveal nucleotide insertions, deletions and rearrangements by virtue of their effect on endonuclease recognition sites in terms of the identity of these sites and their position. The alleles produced by such changes constitute restriction fragment length polymorphisms (RFLP's).

Of an estimated 10^7 such RFLP's in the human genome[10] (1 bp variation every 200 bp) it might be confidently expected that some represent such DNA variation as affects the regulation of gene expression or the nature of the gene product; alternatively they may, by virtue of their proximity on the chromosome, constitute genetic linkage markers for such aetiological loci. Clearly the majority of informative RFLP's will fall into the latter category and provide the commonest source of gene markers.

Linkage markers

Given then that most RFLP's do not represent aetiological mutations and consequently are of use only as linkage markers, it is of great interest to quantify just how well any given RFLP or set of RFLP's (for obviously a marker locus can be expanded to include as many polymorphic sites as can be found at any gene locus) marks possible mutations. Such an index of usefulness would have two components, firstly a measure of how close, in terms of possible recombination, the marker and aetiological mutation are to each other and secondly, a measure of how unique the marker locus is. The first index is variable and depends on the aetiological mutation and the marker locus chosen, it cannot be calculated but must be sought experimentally. The second index, however, is a constant for any RFLP or group of RFLP's and may be calculated from a knowledge of the respective allelic frequencies at the marker locus. Particularly in terms of family studies it is important to maximize this index to achieve maximum resolution of any disease co-segregation. This index has been termed the polymorphism information content value (PIC value).[11]

Informativeness in this context may be described as being the probability that knowing the genotype of an offspring of a parent carrying the aetiological disease mutation would allow deduction of the parental genotype at the marker locus. Given a knowledge of the allelic frequencies at the marker locus such a probability can be calculated for any possible parental mating, in regard to the marker locus and this may then be multiplied by the likelihood of that mating. The resulting probability can then be summed with those from all other possible matings to give the PIC value for that marker locus. Obviously the PIC value may range from zero, marker universally present, to unity, marker individually unique, but in general terms any value greater than 0.25 indicates a usable marker and greater than 0.5 an excellent marker.

Calculation of the PIC value for a locus can be performed by a simple formula as shown below;

$$PIC = 1 - (\Sigma_{n-1}^{n} . P_i^2) - \Sigma_{i=1}^{n-1} . \Sigma_{j=i+1}^{n} . 2P_i^2 P_j^2$$

Where A_i represents the marker locus, with n alleles of frequency p_i.

PIC values can give a good idea of the basic value of a marker locus in a search for linkage with any possible aetiological mutation (but it does not allow for consideration of the complexities of multigenerational studies). In studying a family with more than two generations the effective informativeness of any locus will be proportionally greater than when used for a simple nuclear family.

PIC values for genes thought to be associated with atherosclerosis are given in Table 1. As can be seen the apolipoprotein AI-CIII-AIV gene cluster is highly polymorphic and has a correspondingly high PIC value.

Table 1 Candidate genes for atherosclerosis

Phenotype	Protein	Chromosomal location	PIC
Lipoproteins	Apolipoproteins		
	AI-CIII-AIV	11q23-24	.73/.36/.55
	E-CI-CII	19q13	.36/.28/.79
	B	2p24-23	.66
Receptors	LDL receptor	19p13	.60
	Remnant receptors	—	—
	Insulin receptor	19p13	.80
Enzymes	LCAT	16q22	—
	Lipoprotein Lipase	8p22	.57
Vessel/wall proteins	Fibronectin	2q34-36	.36
	Collagen	17q21-22	.43
Growth Factors	PDGF B	22q12-13	.37
	PDGF A	7p21-p22 or 7q11-12	—
	Epidermal GF	—	—
	Insulin	11p15	.57
Coagulation factors	Fibrinogen A	4q28	.34
	Fibrinogen B	4q28	.28
	Prothrombin	—	—
	Factor VII	13q34	—

Other methods for detection of nucleotide variation

Endonucleases do not represent the sole method for identifying nucleotide variation and although they still provide a reliable and popular method of searching for disease association other more refined techniques are increasingly being used. This is especially relevant as of course some sequence changes do not manifest themselves as changes in endonuclease recognition sites or major positional changes of these sites and are thus beyond the limitations of RFLP studies. Of particular importance amongst these kind of changes are:

Variable number of tandem repeats

These are short repeating sequences usually composed of two nucleotides that are tandemly replicated up to variable lengths of between 15–150 bases. In the human genome there are 50 000 interspersed (CA)n in blocks which if randomly scattered would place them about every 30–60 kb. Like larger hypervariable sequences, they will constitute very useful genetic markers because of length heterogeneity and their large numbers scattered randomly throughout the genome.

If the use of hypervariable polymorphisms were to positively identify a locus incriminated in the inheritance of atherosclerosis, the task would then be to locate the responsible mutation and this might be achieved with one of two approaches;

(a) *Polymerase chain reaction and sequencing*. Defined regions of the incriminated candidate gene (such as single exons or regulatory sequences) may be amplified with the polymerase chain reaction.[12] The products are then digested with restriction enzymes and the fragments analysed directly by gel electrophoresis for differences between cases and controls. Alternatively specific nucleotide changes can be determined by hybridization with end-labelled oligonucleotides complementary to either the common or postulated variant sequences. Amplified DNA is immobilized onto nitrocellulose membranes using a 'slot-blot apparatus' and then hybridized with the 5′ end-labelled oligonucleotide probes and visualized with autoradiography.

(b) *Denaturing gradient gel electrophoresis*. This gel system separates DNA fragments according to their melting properties. When a DNA fragment is electrophoresed through a linearly increasing gradient of denaturants the fragment remains double-stranded until it reaches the concentration of denaturants equivalent to a melting temperature that would cause the lower melting point domains of the fragment to separate. At this stage, disruption of the hybrid molecule caused by a partial melting causes drastic changes in mobility through the gel. A single nucleotide change in a fragment of up to 500 base pairs can be detected by these means. Attachment of a GC-rich sequence of up to 40 nucleotides (a GC clamp) enables minor differences in the higher temperature domains to be detected, allowing the technique to resolve close to 100% of single base-pair differences.

(c) *Base pair specific chemical cleavage.* Chemical cleavage of base pair mismatches in 'patient-native' DNA heteroduplexes formed from polymerase amplified portions of DNA with either hydroxylamine or osmium tetroxide followed by incubation with piperidine may also be used to provide a rapid screening technique for any type of sequence mutation in genes of known wild type sequence.[13]

USING GENE MARKERS

Gene markers may be analysed either in terms of their transmission in pedigrees in which the disease segregates or by their association with a disease phenotype in a given population as compared to another control population. In terms of the application to atherosclerosis it is this latter approach that has proved most informative in the past, almost to the exclusion of the former. This is undoubtedly due to the lack of suitable pedigrees for a disease which takes many years to develop, is difficult to diagnose clinically and severely curtails life expectancy after presentation. In practice it is rare to be presented with more than a single generation with the phenotypic disease. The future of family studies would seem more promising, however, with the development of the powerful single generation analysis afforded by sib-pair comparisons.[14]

This technique at its most simple examines the concordancy of any particular genetic marker between sibs with the phenotypic disease. Such concordancy should be raised above the expected value by the act of selection for the disease if this is genetically linked to the marker under study. Background concordancy rates (the so called 'mean of descent') can be ascertained either by calculation from allelic frequencies in the population or by measuring concordancy in other non-selected sib-pairs ideally from the affected families as this overcomes the objection that population allelic frequency poorly estimates the parental frequency in that the parents of sibs with the phenotypic disease are *ipse facto* a selected population. Disposing of this objection allows the null hypothesis to be simpler and centered solely on sib-pair concordancy. It is worth emphasizing that unselected sib-pairs are exactly that, no decision as to the presence or absence of the phenotypic disease should or needs be made.

The drawback of such an approach lies solely in the number of sib-pairs required, which is dependent on the degree of polymorphism at the marker chosen and its degree of proximity to an aetiological locus. For a uniquely polymorphic site (or to put it another way, one with 'polymorphism information content' (PIC) of unity) such as the HLA system, all sib-pairs analysed contribute to a result and only low numbers of pairs are needed but the lower the PIC value the higher the chance that in a given sib-pair the chosen marker will not allow the identification of parental phase and consequently cannot be used to contribute to an interpretable result. This greatly increases the number of pairs required to reach a conclusion. It is in view of this that the ideal markers for the approach prove to be hypervariable regions and variable number tandem repeats which are highly polymorphic and spread throughout the genome.

However, notwithstanding the possible future importance of such analysis, the bulk of current knowledge is based on the foundation of a relatively simple population association approach.

CANDIDATE GENES

Which of the 1.4 million potential genes in the human genome are likely to be involved in the pathogenesis of atherosclerosis? Answers to this question can be achieved by two differing but complementary approaches.

Candidate gene targeting

This assumes that those genes known to code for a protein suspected to be involved in the atherogenic process are the genes most worthy of study. For example a candidate gene approach may involve concentrating on a gene producing a protein known to be central to lipid metabolism, such as apolipoprotein B, searching for RFLP's at or around that locus and then, having identified such polymorphisms, proceeding to see if the relative frequencies are altered between patient and control groups in various racially differentiated populations.

Complete genomic mapping

This is based on the production of complementary DNA and genomic DNA libraries, which allow the isolation of random unique DNA fragments with regular spacings along each and every chromosome and the subsequent use of these gene fragments as hybridization probes to detect RFLP's. Pedigree studies are then pursued and should any association be found, the gene fragment can be mapped to the genome. Then by 'walking' along the chromosomal segment with further probes[15] one might expect to be able to pinpoint the aetiological mutation involved.

Of most molecular genetic work done on multifactorial disease to date the overwhelming preponderance has been of this targeted type approach and this holds true for atherosclerosis.

The identification of candidate genes uses as its source the wealth of data regarding proteins that having been identified and characterized and are thought to be implicated in the development or onset of atherosclerosis. Examples of such include the apolipoproteins, the insulin gene and the LDL receptor gene and many others. A list of candidate genes is presented in Table 1. Some of these showing protein polymorphisms have already been studied with regard to associations with premature atherosclerosis.

PROTEIN POLYMORPHISMS

Apolipoprotein E

This polymorphism is controlled by three common alleles E2, E3 and E4 (Table 2) and by a series of rare alleles more often found in patients with type III hyperlipidaemia.[16] The three common protein isoforms of Apo E2, Apo E3 and Apo E4 differ by amino acid replacements in two positions of the protein sequence and by their functional properties. For example Apo E2 (arg-158 to cys) is defective in binding to lipoprotein receptors and this may, when homozygous, constitute a genetic component of type III hyperlipidaemia. Most Apo E2/2 subjects, however, never develop hyperlipidaemia, but on the contrary have subnormal levels of plasma

Table 2 The apoliprotein E polymorphism and gene frequencies (data modified after Utermann[63])

Apo E alleles	Protein	Polymorphism
E2	Apo E2 (arg-158 to cys)	Receptor binding activity <2% of apo E3
E3	Apo E3	—
E4	Apo E4 (Cys-112 to arg)	Enhanced in vivo catabolism

Frequencies of alleles of Apo E

Apo E alleles	Finns (n = 408)	Germans (n = 1031)	Japanese (n = 319)
E2	0.029	0.077	0.081
E3	0.750	0.773	0.849
E4	0.221	0.150	0.067

cholesterol (mean effect of -14.2 mg/dl) due to reduced concentrations of LDL. Conversely subjects with the E4 allele have a mean raised plasma cholesterol of $+7$ mg/dl. The ApoE gene locus has been claimed to account for 4% and 20% of the phenotypic variance of cholesterol and ApoE concentrations respectively in German populations.[17] This is reflected in different levels of LDL cholesterol (but not total cholesterol) between survivors of myocardial infarction who are ApoE 3/4 or ApoE 2/3 of $199.2 \pm 45(80)$ mg/dl and $163 \pm 50(33)$ mg/dl respectively.[18] Another study[19] examined the polymorphism at the ApoE locus in relation to risk of early coronary artery disease. The isotype frequencies were determined in a random sample of 400 persons aged 45–60 years in North East Scotland and compared to those of a group of survivors of myocardial infarction aged under 56 collected from diverse sources (a source of possible confusion in interpreting the results). E4/3 was more frequently seen in the coronary artery disease group at the expense of E3/2 (0.32 v. 0.25 and 0.075 v. 0.127, $P < 0.05$) and for survivors aged under sixty this heterogeneity was even more marked ($P < 0.01$).

Comparison of the average age of first myocardial infarction in male survivors also suggested that this may have occurred earlier in those of genotype E4/3 (E3/3 53.93 ± 0.68, E4/3 51.20 ± 0.98 and E3/2 53.21 ± 2.02; E4/3 differs significantly $P < 0.05$). This does raise the possibility that the E4 allele may play a role in myocardial infarction but the frequency of alternative genotypes amongst survivors of myocardial infarction will be influenced by the risk of coronary death if this too varies between genotypes. Other studies have previously noted an increase in the frequency of the E3/2 genotype in selected patients with premature coronary artery disease so such an explanation cannot be entirely ruled out.

Lp(a) lipoprotein

The Lp(a) antigen was first described by Berg[20] in 1963. Recent studies have established that Lp(a) represents an LDL particle in which Apo B100 is disulphide-linked to Apo(a), a glycoprotein with a striking homology to plasminogen.[21] Lp(a) is heterogeneous in size and density, in part because Apo(a) varies in mass between 280 000 and 700 000 daltons. This heterogeneity is due to the Apo(a) gene (on chromosome 6) coding for various polypeptides with different numbers of Kringle 4 domains.[22] At least seven Apo(a) isoforms have been described and their size appears to be inversely correlated to Lp(a) concentration. The cysteine involved in the disulphide linkage of Apo(a) to Apo B is probably located in Kringle 36 of Apo(a). Where this linkage occurs and the steps in the cellular assembly of Lp(a) are currently unclear, but the liver has been established as a major source of Lp(a). The molecular heterogeneity of Lp(a) and its close similarity to plasminogen has made its assay difficult for clinical studies and although the most reliable assay for Lp(a) has been shown to be ELISA, which appears to be free of interference by plasminogen, not many clinical studies have used this. Other methods have, however, shown that in case-control studies Lp(a) levels of more than approximately 20 mg/dl are associated with an increased risk of myocardial or cerebrovascular infarction.[23–25] A recent study has shown that individuals with a plasma concentration of greater than 25 mg/dl exhibit a twofold higher risk of myocardial infarction than controls.[25]

As to the mechanism of this association the similarity between Apo(a) and plasminogen raises the possibility that Lp(a) might play a role in the fibrinolytic system by blocking the action of plasminogen and predisposing to thrombosis. The principal proteolytic enzymes involved in fibrinolysis are plasmin and its zymogen plasminogen. The latter may become adsorbed and concentrated onto fibrin together with its activator during the conversion of fibrinogen to fibrin so that the formation of plasmin occurs directly on its substrate. The interaction of plasminogen and fibrin is mediated by two types of binding sites located on Kringle 1 and 4. Since Apo(a) contains only the weaker binding Kringle 4 domains, Lp(a) binds fibrin only weakly if at all. It may, however, interfere with plasminogen function and have a thrombogenic effect. The issue is complicated by the various sizes of the isoforms of Apo(a) as well as their concentration in plasma; and it is possible that the larger isoforms may have different effects on plasminogen from the smaller ones. All isoforms of Lp(a) therefore need to be studied before their thrombotic potential can be established. There are therefore many unresolved issues concerning Lp(a), in particular its physiological role, its regulation, synthesis and catabolism. The establishment of normal plasma levels regarding mass and isoforms need to be clarified before its role in atherogenesis can be fully evaluated.

GENETIC POLYMORPHISMS

Complementary to this data regarding protein polymorphism there also exists a great deal of data regarding genetic polymorphism and atherosclerosis. Some of the most promising of which is reviewed below.

The apolipoprotein AI-CIII-AIV gene cluster

The above three genes are clustered on the long arm of chromosome 11 on a DNA segment of approximately 4 kb.[26,27] The organization of the cluster is presented in Figure 2 and shows the following features;

1. The Apo CIII gene is transcribed in the opposite direction to the Apo AI and AIV genes despite being within 3 kb of the 3′ end of the Apo Al gene.

2. More than 9 restriction enzyme dimorphisms along the length of this part of the genome have been described occurring within introns, exons, intergenic sequences and in flanking sequences.[28,29] Many population studies have been performed (from the United Kingdom, West Germany, United States and Japan), examining the frequencies of allelic variants at these restriction sites to see if any associate with premature atherosclerosis. The results are as follows:

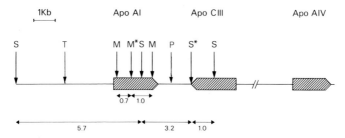

Fig. 2. The organization of the apolipoprotein AI-CIII-AIV gene cluster on the long arm of chromosome 11. The map shows polymorphic restriction sites for the enzymes: S = Sst I, X = I, P = Pst I, M = Msp I and T = Taq I.

Table 3 The Apo CIII mutation: frequency of the S2 allelic variants in control and patient populations

		Allelic frequencies		
	n	S1	S2	References
Control groups				
Random medical outpatients	37	0.96	0.04	Rees et al.[28]
Health screen clinics				
Samples 1	42	1.0	0.00	Rees et al.[31]
2	74	0.98	0.02	Ferns et al.[30]
3	56	0.98	0.02	O'Connor et al.[59]
Normal coronary arteries	68	0.97	0.03	Rees et al.[31]
Random medical outpatients	35	0.99	0.01	Trembath et al.[60]
Normolipidaemic controls	71	0.94	0.06	Kessling et al.[61]
Random normals	101	0.94	0.06	Deeb et al.[35]
Controls	66	0.98	0.02	Hegele et al.[36]
Patient groups				
Hyperlipidaemic (iv/v)	28	0.80	0.20	Rees et al.[28]
Survivors of MI	48	0.88	0.12	Ferns et al.[30]
Coronary atheroma	61	0.89	0.11	Rees et al.[31]
Peripheral atheroma	49	0.88	0.12	O'Connor et al.[59]
Diabetic survivors of MI	47	0.86	0.14	Trembath et al.[60]
Hyperlipidaemia with gout	22	0.88	0.12	Ferns et al.[62]
Coronary heart disease	140	0.88	0.12	Deeb et al.[35]
Survivors of MI	66	0.96	0.04	Hegele et al.[36]

United Kingdom

Two groups of patients have been studied, young survivors of myocardial infarction[30] and patients with coronary or extracoronary atheroma proven by angiography.[31] In the former group, the frequency of an uncommon allele (the S2 allele) at the Sst I restriction site in the fourth exon of the Apo C-III gene was approximately 4% in healthy controls (n = 47) compared to 21% in young survivors of myocardial infarction (n = 48). When other restriction site polymorphisms (Msp I and Pst I) were included in the analysis thereby constructing DNA haplotypes,[32] it was found that one particular haplotype containing the uncommon allele at the Msp I and Sst I sites was increased from 2% in normolipaemic controls (n = 48) to 21% in survivors of myocardial infarction (n = 47) giving a relative incidence of 12.7 ($P < 0.01$). However because of tight linkage disequilibrium between the alleles studied, it was not possible to identify haplotypes associated with any greater risk of premature atherosclerosis than when the Sst I polymorphism was considered in isolation. In a study from Edinburgh,[35] the frequency of the S2 allele combined with rare alleles at Xmn I, Pst I and Msp I sites showed an increased frequency in patients with coronary artery disease who also had a positive family history compared to those without a family history (relative prevalence 3.34; $P < 0.0005$).

Such results must be interpreted with caution however. A common disease such as premature atherosclerosis may be particularly heterogeneous with different polygenic determinants operating in different geographical localities. The patient groups must be clearly defined and made as homogeneous as possible with regard to racial and geographical origins and clinical diagnostic features.

Since atherosclerosis may have a very variable age of onset before the diagnosis can be established with certainty, the control groups may contain individuals who will go on to develop atherosclerosis at a later age. For example, the frequency of the S2 allele in healthy subjects from Scotland was reported as 18% (n = 64) compared to 4% in London (n = 47). This may represent a real difference in allelic frequencies or simply be due to differences in exclusion criteria for constituting a healthy group (i.e. presence or absence of hyperlipidaemia, diabetes or a family history of coronary heart disease, etc). It is of interest that the S2 allele frequencies in healthy Caucasian groups from Boston and Seattle are 5% (n = 66) and 6% (n = 101) respectively. This suggests that it is important to establish strict inclusion criteria for healthy control groups.

Other studies[31] have used coronary arteriography to define the presence of atherosclerosis. In one report the frequency of the S2 allele was 22% in patients with severe obstructive coronary atherosclerosis (n = 61) compared to 6% in patients with minimal disease (n = 68, $P < 0.02$).

United States

Studies from Boston, Seattle and New York have been reported. In the first[33] Caucasian patients (n = 88) with severe coronary artery disease were compared to a Framingham control population (n = 64) matched for ethnic origin and with other clinical criteria carefully standardized. The frequency of an uncommon allele revealed by the enzyme Pst I at a restriction site 34 base pairs 3' to the Apo AI gene was 32% in patients compared to 4% in matched controls ($P < 0.01$) and 3% in 30 subjects with no angiographic evidence for coronary artery disease, giving a relative risk of coronary artery disease in individuals possessing the P2 allele of at least 10. The same rare allele was found at increased frequency in subjects with familial hypoalphalipoproteinaemia and the effect of this mutation on atherosclerosis may be mediated by lowering the levels of plasma HDL. Frequencies of alleles at other polymorphic sites at this locus were not reported. However, in a study from Seattle,[35] frequencies of alleles revealed by Pst I and Sst I restriction enzymes were compared in patient groups with coronary artery disease proven by arteriography (n = 140) and random 'normals' (inclusion criteria not stated). No differences were observed in the frequency of the P2 allele, however the frequency in the control group was 10% which is three times that of the Boston Study. Clearly this will tend to minimize differences between patients and 'normals.' With regard to the S2 allele, frequencies were 6% (n = 101) and 12% (n = 140) in 'normals' and patients respectively ($P < 0.05$).

In a third study,[36] survivors of myocardial infarcts were examined for allelic frequencies at four polymorphic sites revealed with the enzymes Xmn I, Msp I, Pst I and Sac I. The only significant difference was observed with the Xmn I polymorphism where the X2 allele frequency was lower in the group (n = 57) at 15% compared to 24% in controls (n = 57, $P < 0.05$). When individuals above the age of 60 years were studied the P2 allele was significantly less frequent in patients (3% versus 21% for controls, $P < 0.02$). The authors concluded that DNA polymorphisms near the Apo AI gene may be significantly associated with myocardial infarction.

West Germany

A study from Munster,[37] examined eight polymorphic sites at the Apo AI-CIII-AIV gene cluster. These included sites for the

enzymes Apa I, Msp I, Pst I, Bam II and Pvu II. Pseudohaplo-
types were constructed from these data for 314 patients suffering
from myocardial infarction before the age of 45 years to compare
to 267 student controls. Frequencies of haplotypes containing the
rare allele of Pvu II at the 5' site of the Apo AI gene showed an
increased frequency from 0.04 to 0.09 ($P < 0.05$). Other pseudo-
haplotype frequencies were not significantly different. This however is probably an underestimate in view of the fact the controls
were students and two decades younger than the patient groups.

Japan

A study was conducted in North Japan in 69 subjects surviving a
myocardial infarction for comparison with 82 healthy controls. No
significant differences were found with the rare alleles of the Sst
I or the Msp I sites; but the control frequencies were very different
from Caucasian frequencies. However, the haplotype S1-M2 was
significantly increased in myocardial infarction survivors compared to controls (24% versus 11%, $P < 0.05$, $X^2 = 4.90$); suggesting that this haplotype may be a marker for a putative atherogenic
allele in the vicinity.

All studies thus far show restriction site polymorphisms that
associate in patient groups with atherosclerosis defined either by
angiography or myocardial infarctions, but the involved sites differ
markedly between populations. This may be expected if the sites
are only acting as linkage markers for an atherogenic allele in the
vicinity. There is also variability amongst studies regarding the
frequencies of allelic variants in control populations and by affecting the comparisons with patient groups may account for some of
the inconsistencies.

However, some tentative conclusions may be drawn. Firstly the
restriction site polymorphisms so far studied, probably arise from
harmless mutations and are not functioning in any way as aetiological determinants. They are possibly background (or neutral) DNA
variants that differ in frequencies amongst Caucasian and other
racial populations in the same way that some HLA antigen frequencies are found to vary amongst different Caucasian populations. The DNA polymorphisms, however, may be acting as
linkage markers for neighbouring atherogenic mutations. There
are at least two possible reasons why different neutral polymorphisms may act as linkage markers within a racial group. Firstly
the putative atherogenic allele may have mutated more than once

in different geographical localities and become associated with different background polymorphisms depending on which chromosome the atherogenic mutation occurred. The background polymorphisms may have since 'hitch-hiked' with the mutated atherogenic allele.

An example of this is the sickle cell mutation in the B-globin gene in West Africans that occurred on a chromosome carrying an Hpa I polymorphism 3 kb downstream from the globin gene.[38] This Hpa I polymorphism acts as a linkage marker for the sickle-cell mutation, but only in West African populations. Another example of this in the field of lipid metabolism is familial hypertriglyceridaemia due to mutations within the Apo C-II gene. In a pedigree study[39] from North Italy and Holland it was observed that different Apo C-II alleles revealed by a Taq I restriction site polymorphism tracked with affected members of each pedigree. In the North Italian family, the affected members were associated with a 3.8 kb allele; in the Dutch family by a 3.5 kb allele. The simplest explanation for these results is the Apo C-II mutation has occurred at least twice on different chromosomes carrying different background polymorphisms at the Taq I restriction site. Subsequent studies of the C-II apolipoproteins have shown the occurrence of different amino-acid replacements amongst pedigrees with this form of familial hypertriglyceridaemia.[40]

The apolipoprotein B gene

The human Apo B gene has been cloned and localised to chromosome 2 in the region of p24.[41,42] It extends over 43 kb containing 29 exons and 28 introns. The distribution of introns is asymmetrical, occurring in the 5'-terminal third of the gene. The complete amino acid sequence (mol wt 54 kD) has been deduced from a cDNA clone.[43] A domain enriched in basic amino acids has been identified as important for the cellular uptake of cholesterol by the LDL receptor pathway. Many restriction site polymorphisms have been observed including those for the enzymes Xba I, Eco RI and Msp I; and several studies have been published on their distribution in patients with coronary artery disease. In one UK study, no significant difference was observed in the allelic frequencies of the Xba I polymorphism at the 3' end of the Apo B gene in 52 survivors of myocardial infarction and 33 healthy controls.[44] This was also observed in a second study in the USA where very similar allele frequencies were reported.[35] However, the latter group found a change in an insertional-deletional polymorphism revealed by the enzyme Msp I that increased in frequency in controls (n = 62) from 0.06 to 0.15 in angiographically proven patients with coronary artery disease (n = 103). This was supported in a further study[45] where the frequency of the insertional polymorphism increased from 0.142 in controls (n = 84) to 0.267 in patients (n = 84). In addition, the latter study reported an increased frequency of the rarer allele of the Xba I polymorphism in patients compared to controls. They concluded that the variant sites producing these polymorphisms were unlikely to be causing abnormalities of apolipoprotein B but may be simply genetic markers in linkage disequilibrium with other atherogenic mutations. None of these restriction site polymorphisms were associated with variation in lipid, lipoprotein or apolipoprotein levels and may therefore be independent risk factors for coronary artery disease. A further study from the UK[46] found that frequencies of the E- allele (Eco RI restriction site absent) and X- allele (Xba I site absent) were both significantly higher in normocholesterolaemic men with coronary artery disease than those without coronary artery disease. But the associations were weak and would require further work to elucidate a more informative locus. It was of interest that none of these studies revealed differences in allelic or genotype frequencies between cases and controls that were of any greater magnitude than those reported for polymorphisms of the Apo AI-CIII-AIV gene cluster. A very large difference in allele frequency between patients and controls would imply that variation at that particular locus is associated with the development of coronary artery disease in a large proportion of cases in the general population. This may be unlikely in view of the heterogeneity of the disease and the random selection of patients in the current studies, without using any particular phenotypic feature, such as age of onset or major associated risk factors.

The inconsistencies between the reports so far published, may arise from variation of allelic frequencies at these polymorphic sites within genetic sub-groups of a population. To interpret subsequent studies, it will be very important to ensure that cases and controls come from the same gene pool, i.e. belong to the same ethnic subgroups and preferably arise from the same geographical locality. Repetition of such studies in other racial subgroups will be important to determine how widespread are these associations and the nature of their possible phenotypic consequences.

Low density lipoprotein receptor gene

The elevation in LDL cholesterol resulting from familial hypercholesterolaemia (FH) is known to lead to premature coronary atherosclerosis and accounts for up to 6% of myocardial infarction occurring before the age of 60 years.[47] The human LDL receptor gene has been cloned and localized to chromosome 19 and consists of 18 exons, 13 of which seem to have marked sequence homology with the genes for the C9 component of complement and for epidermal growth factor.[48,49]

RFLP studies for this locus have demonstrated segregation with FH in isolated families[50,51] and various categories of gene defects have been shown to be responsible for LDL receptor failure in certain small and well characterized hypercholesterolaemic groups.[52-54] In certain instances homozygous FH patients can be shown to possess either similar or dissimilar gene defects on each parental chromosome. Work within populations such as the Quebecois French-Canadians and the Lebanese, with their marked founder and inbreeding effects, has allowed full characterization of LDL receptor defects within these populations which has in turn allowed functional identification of certain exons of the gene. In many ways the work on this gene provides a prototype for this kind of study in general and its importance in this respect extends far beyond the direct implications for familial hypercholesterolaemia.

Lipoprotein lipase gene

The gene for human lipoprotein lipase has been cloned and localised to chromosome 8 in the region p22[55] and its gene product is known to play a central role in lipid catabolism, catalysing the rate limiting and flow generating step in the removal of triglyceride-rich lipoprotein particles from the plasma. The configuration of the gene is presented in Figure 3. RFLP studies at this locus have yielded a promising association between the Hind III restriction site found in the 3' untranslated portion of the gene beyond exon 10 and hypertriglyceridaemia[56,57] in both a United Kingdom Caucasian and Japanese population; and the Pvu II site of exon 6 has been implicated in an associative contribution to the population range of triglyceride level. The Hind III polymorphism would also appear to be linked to angiographically defined premature coronary artery disease in a United Kingdom Caucasian population.[58]

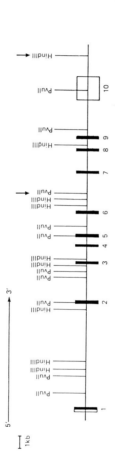

Fig. 3. A map of the lipoprotein lipase gene on chromosome 8p22 showing restriction sites. Polymorphic restriction sites are arrowed.

Although triglyceride rich lipoproteins may not be an independent risk factor from HDL for the development of coronary heart disease, the close metabolic inter-relationship between HDL and triglyceride transport makes it very difficult to evaluate their distinct independent roles. Hypertriglyceridaemia, with low HDL, may constitute a risk factor for the development of coronary atherosclerosis and the association of an RFLP allele at the lipoprotein lipase gene locus with both hypertriglyceridaemia and coronary artery disease suggests that this gene might be a causal genetic determinant of atherosclerosis, although it is too early to be certain at this stage and confirmatory work must be awaited.

CONCLUSION

All the studies here are subsequent on a major new ability of recombinant DNA technology. That is the ability to detect genetic variants that do not necessarily give rise to mutant proteins. Such variants may of course be aetiological for a disease process or act as linkage markers for another variant that is.

The application of such techniques to atherosclerosis has primarily been centered on a search for RFLP linkage markers in particular candidate genes and much interesting data has arisen from such work. It is to be hoped that the future will bring both a wider application of these techniques together with the use of other more advanced analyses to characterize the underlying aetiological mutations. If there is to be progress made in our understanding of atherosclerosis recombinant DNA technology appears to offer the tools we need to achieve it. A rapid expansion in the understanding of coronary artery disease and other polygenic diseases is surely imminent.

ACKNOWLEDGEMENTS

The authors would like to thank the British Heart Foundation together with the Medical Research Council (UK) and the British Diabetic Association for financial assistance.

REFERENCES

1 Yater WM, Traum AH, Brown WG, Fitzgerald RP, Geisler MA, Wilcox BB. Coronary artery disease in men eighteen to thirty-nine years of age. Am Heart J 1948; 36: 334
2 Slack J, Evans KA. The increased risk of death from ischaemic heart disease

in first-degree relatives of 121 men and 96 women with ischaemic heart disease. J Med Genet 1966; 3: 239

3 Rissanen AM. Familial aggregation of coronary heart disease in a high incidence area (North Karelia, Finland). Br Heart J 1979; 42: 294

4 Nora JJ, Lortscher RM, Spangler RD, Kimberling WJ. Genetic epidemiology study of early onset ischaemic heart disease. Circulation 1980: 61: 503

5 Berg K. Twin studies of coronary heart disease and its risk factors. Acta Genet Med Gemell 1984; 33: 349

6 Berg K. Genetics of coronary heart disease. Prog Med Genet 1983; 5: 36

7 Kannel WB, Shurltleft D. The natural history of arteriosclerosis obliterans. In: Peripheral vascular disease, Vol. 3. Philadelphia: Davis, 1971: p. 38

8 Widmer LK, Stakelin HB, Nissen C, Silva V. Venen und Arterienkrankheiten, und koronare Herzkrankheit bei Berufst tigen. Bern: Huber, 1981

9 Southern EM. Detection of specific sequences among DNA fragments separated by gel electrophoresis. J Mol Biol 1975; 98: 503

10 Jeffrey AJ, DNA sequence variants in the gamma, alpha, delta and beta globin genes of man. Cell 1979; 18: 1

11 Botstein D, White RL, Skolnick M, Davis RW. Construction of a genetic linkage map in man using restriction fragment length polymorphisms. Am J Hum Genet 1980; 32: 314

12 Saiki RK, Gelfand DH, Stoffel S et al. Primer directed enzymatic amplification of DNA with a thermostable DNA polymerase. Science 1988; 239: 487

13 Cotton RGH, Rodrigues NR, Campbell RD. Reactivity of cytosine and thymine in single base pair mismatches and its application to the study of mutations. Proc Natl Acad Sci USA 1988; 85: 4397

14 Galton DJ, Oelbaum RS, Chamberlain JC. The use of restriction site polymorphisms in the analysis of common polygenic disease. In: Proceedings of the British Nutrition Foundation. London. (In press)

15 Wiessman SM. Molecular genetic techniques for mapping the human genome. Mol Biol Med 1987; 4: 133

16 Utermann G, Hees M, Steinmetz A. Polymorphism of apolipoprotein E and occurrence of dysbetalipoproteinaemia in man. Nature 1977: 269: 604

17 Utermann G. Apolipoproteins, quantitative lipoprotein traits and multifactorial hyperlipidaemia. CIBA Foundation Symposium 1987; 130: 52

18 Assmann G, Schulte H, Funke H, Schmitz G, Robenck H. High density lipoproteins and atherosclerosis. Atherosclerosis 1989; 75: 341

19 Cumming AN, Robertson FW. Polymorphism at the apolipoprotein E locus in relation to risk of coronary disease. Clin Genet 1984; 25: 310

20 Berg K. A new serum type system in man: The Lp- system. Acta Pathol Microbiol Scand 1963; 59: 369

21 Scanu AM, Fless GM. Lp(a): a lipoprotein particle with atherogenic and thrombogenic potential. Atherosclerosis 1989; 75: 189

22 McClean JW, Tomlinson JE, Kuang WJ et al. cDNA sequence of human apolipoprotein (a) is homologous to plasminogen. Nature 1987; 330: 132

23 Kostner GM, Avogaro P, Gazzolato G, Marth E, Bittolobon G. Lipoprotein Lp(a) and the risk of myocardial infarction. Atherosclerosis 1981; 38: 51

24 Armstrong VW, Walli AK, Seidel D. Isolation, characterization and uptake in human fibroblasts of an apo(a) free lipoprotein obtained on reduction of lipoprotein (a). J Lipid Res 1985; 26: 1314

25 Kostner GM. Lipoprotein Lp(a) and HMG-CoA reductase inhibitors. Atherosclerosis 1989; 75: 405

26 Karathanasis SK, Zannis VI, Breslow JL. Linkage of human apolipoprotein AI and CIII genes. Nature 1983; 304: 371

27 Karathanasis SK. Apolipoprotein multigene family: tandem organisation of human apolipoprotein AI, CIII and AIV genes. Proc Natl Acad Sci USA 1985; 82: 6374

28 Rees A, Stocks J, Shoulders CC, Galton DJ, Baralle FE. DNA polymorphism adjacent to the human apoprotein AI gene in relation to hypertriglyceridaemia. Lancet 1983; i: 444

29 Seilhammer JJ, Protter AA, Frossard P, Levy-Wilson B. Isolation and DNA sequence of full length cDNA of the entire gene for human apolipoprotein AI. Discovery of a new polymorphism. DNA 1984; 3: 309

30 Ferns GAA, Stocks J, Ritchie C, Galton DJ. Genetic polymorphisms of apolipoprotein C-III and insulin in survivors of myocardial infarction. Lancet 1985; i: 300

31 Rees A, Jowett NI, Williams LG et al. DNA polymorphisms flanking the insulin and apolipoprotein CIII genes and atherosclerosis. Atherosclerosis 1985; 58: 269

32 Ferns GAA, Galton DJ. Haplotypes of the human apoprotein AI-CIII-AIV gene cluster in coronary atherosclerosis. Hum Genet 1986; 73: 245

33 Price WH, Morris SW, Kitchin AH, Wenham PR, Burgon PRS, Donald PM. DNA restriction fragment length polymorphisms as markers of familial coronary artery disease. Lancet 1989; i: 407

34 Ordovas JM, Schaeffer EJ, Salem D et al. Apolipoprotein A-I gene polymorphism associated with premature coronary artery disease and familial hypoalphalipoproteinaemia. N Engl J Med 1986; 314: 671

35 Deeb S, Failor A, Brown BG, Brunzell JD, Albers JJ, Motulsky AG. Molecular genetics of apolipoproteins and coronary heart disease. Cold Spring Harb Symp Quant Biol 1987; 51: 403

36 Hegele RA, Herbert PN, Blum CB, Buring JE, Hennekens CH, Breslow JL. Apolipoprotein AI and AII gene DNA polymorphisms and myocardial infarction. (personal communication)

37 Assman G, Schulte H, Funke H, Schmitz G, Robenck H. High Density Lipoproteins and atherosclerosis. Atherosclerosis 1989; 75: 34

38 Yan YW, Dozy AM. Polymorphisms of DNA sequence adjacent to human beta globin structural gene: relation to sickle cell mutation. Proc Natl Acad Sci USA 1978; 75: 5631

39 Humphries SE, Williams LG, Stalhenhoef AF et al. Familial apolipoprotein CII deficiency: a preliminary analysis of the gene defect in two pedigrees. Hum Genet 1984; 65: 151

40 Baggio G, Gabeli, C, Manzato E et al. Apo CII Padova: a new apoprotein variant in two patients with apo CII deficiency syndrome. In: Sirtori C, Franceschini G, eds. Proceedings of NATO Advanced Research Workshop on Apolipoprotein Mutants, 1986: p. 203

41 Knott JJ, Rall SC Jnr, Innerarity TL et al. Human apolipoprotein B: structure of carboxy-terminal domain sites of gene expression and chromosomal localisation. Science 1985; 230: 37

42 Shoulders CC, Myant N, Sidoli A, Rodriguez JC, Cortese C, Baralle FE. Molecular cloning of human LDL apolipoprotein B cDNA. Atherosclerosis 1985; 58: 277

43 Knott TJ, Pease RJ, Powell LM et al. Human apolipoprotein B: complete cDNA sequence and identification of domains of the protein. Nature 1986; 323: 734

44 Ferns GAA, Galton DJ. Frequency of the Xba I polymorphisms of the apolipoprotein B gene in myocardial infarct survivors. Lancet 1986; ii: 572

45 Hegele RA, Huang LS, Herbert PN et al. Apolipoprotein B-gene DNA polymorphisms associated with myocardial infarction. N Engl J Med 1986; 315: 1509

46 Myant NB, Gallagher J, Barbir M, Thompson GR, Wile D, Humphries SE. RFLP's in the apo B gene in relation to coronary artery disease. Atherosclerosis 1989; 77: 193

47 Goldstein JL, Hazzard WR, Schrott HG. Hyperlipidaemia in coronary heart

disease. Lipid levels in 500 survivors of myocardial infarction. J Clin Invest 1973; 52: 1533

48 Sudhof TC, Goldstein JL, Brown MS, Russell DW. The LDL receptor gene: a mosaic of exons shared with different proteins. Science 1985; 228: 815

49 Francke U, Brown MS, Goldstein JL. Assignment of the human gene for the low density lipoprotein receptor to chromosome 19: Synteny of a receptor, a ligand and a genetic disease. Proc Natl Acad Sci USA 1986; 81: 2826

50 Humphries SE, Hortshemke B, Seed M et al. A common DNA polymorphism of the LDL receptor gene and its use in diagnosis. Lancet 1985; i: 1003

51 Leppert MF, Hasstedt SJ, Holm T et al. A DNA probe for the LDL receptor gene is tightly linked to hypercholesterolaemia in a pedigree with early coronary disease. Am J Hum Genet 1986; 39: 300

52 Tolleshaug H, Hobgood KK, Brown MS, Goldstein JL. The LDL receptor locus in familial hypercholesterolaemia. Multiple mutations disrupt the transport and processing of a membrane receptor. Cell 1983; 32: 941

53 Lehrmann MA, Schneider WJ, Sudhof TC, Brown MS, Goldstein JL, Russell DW. Mutation in LDL receptor: Alu-Alu recombination deletes exons encoding transmembrane and cytoplasmic domains. Science 1985; 227: 140

54 Horsthemke B, Kessling AM, Seed M, Wynn V, Williamson R, Humphries SE. Identification of a deletion in the low density lipoprotein (LDL) receptor gene in a patient with familial hypercholesterolaemia. Hum Genet 1985; 71: 75

55 Sparkes RS, Zollman S, Klisak I et al. Human genes involved in lipolysis of plasma lipoproteins: mapping of loci for lipoprotein lipase to 8p22 and hepatic lipase to 15q21. Genomics 1987; 1(ii): 6793

56 Chamberlain JC, Thorn JA, Oka K, Galton DJ, Stocks J. DNA polymorphisms at the lipoprotein lipase gene: associations in normal and hypertriglyceridaemic subjects. Atherosclerosis 1989; 79: 85

57 Chamberlain JC, Thorn JA, Stocks J, Galton DJ. Genetic variants at the lipoprotein lipase gene locus associate with triglyceride levels in normal and hypertriglyceridaemic subjects. Atherosclerosis 1989; 75: 209

58 Thorn J, Chamberlain JC, Stocks J, Galton DJ. RFLP's at the lipoprotein lipase and hepatic lipase gene loci in coronary atherosclerosis. Atherosclerosis 1989; 79: 94

59 O'Connor G, Stocks J, Lumley J, Galton DJ. A DNA polymorphism of the apolipoprotein CIII gene in extracoronary atherosclerosis. Clin Sci 1988; 74: 289

60 Trembath RC, Thomas DJ, Hendra T, Yudkin J, Galton DJ. A DNA polymorphism of the apo AI-CIII-AIV gene cluster associates with coronary heart disease in non-insulin dependent diabetes. Br Med J 1987; 294: 1577

61 Kessling AM, Horsthemke B, Humphries SE. A study of DNA polymorphisms around the human apoliprotein AI gene in hyperlipidaemic and normal subjects. Clin Genet 1985; 28: 296

62 Ferns GAA, Lanham J, Galton DJ. Polymorphisms of the apolipoprotein AI-CIII gene cluster in subjects with hypertriglyceridaemia associates with primary gout. Hum Genet 1987; 75: 121

63 Utermann G. Apolipoproteins, quantitative lipoprotein traits and multifactorial hyperlipidaemia. CIBA Foundation Symposium 1987; 130: 52

British Medical Bulletin (1990) Vol. 46, No. 4, pp. 941–959

DNA based diagnostic tests: Recombinant DNA and cardiovascular disease risk factors

A Sidoli
S Galliani
F E Baralle★
Department of Biotechnology, Istituto Sieroterapico Milanese S. Belfanti, Milan, and ★International Centre for Genetic Engineering and Biotechnology, Padriciano 99, Trieste, Italy

Advances in molecular biology and medical biotechnology are continuously creating exciting possibilities for DNA based diagnostics. It is now possible by simple procedures to detect polymorphic DNA markers, structural variants and regulatory mutants of human genes, allowing detailed genotyping of patients. The innovative combination of immunoenzymatic techniques, monoclonal antibodies and recombinant tracer proteins, results in new DNA based tests for the determination of important biochemical parameters, in order to define more precisely the phenotype and hence assess the individual risk.

   The application of these technologies to the analysis of dyslipidemias, atherosclerosis and cardiovascular diseases may not only lead to a better understanding of the molecular and genetic basis of these pathologies, but also to their early recognition and better management.

Our understanding of genetic disorders have made prodigious progress after the advent of recombinant DNA techniques.[1-3] The specific genes involved in many inherited diseases have now been cloned and sequenced and the causative mutations identified. Other genetic disorders, for which the genes are not yet isolated,

0007–1420/90/0046–0941/$10.00

have been mapped to particular chromosomal locations, thus allowing predictive testing with linked DNA markers.

We shall discuss the different approaches provided by the new DNA technology for the analysis and diagnosis of the genetic factors involved in the pathogenesis of dyslipidemias, atherosclerosis and cardiovascular diseases (CVD). The hyperlipidemias (with hypertension, diabetes mellitus, and cigarette smoking) are among the major risk factors for the development of atheroma. Familial studies have shown that there is a strong inherited component in CVD and dyslipidemic states, but only in the minority of patients the condition is transmitted as a monogenic defect. In most cases a polygenic condition (i.e. genes involved in lipid metabolism, arterial blood pressure and arterial wall integrity) is interacting with environmental factors in the aetiology of the disease.[4] Conventionally, hyperlipidemias have been studied by looking for abnormalities in protein structure and function in clearly defined disorders. For example the defect in familiar hypercholesterolemia was traced to the cell surface receptor that normally controls the degradation of LDL. This approach, however, is limited because it requires that the gene products differ in size, charge or activity in order to be recognized. Perhaps therefore it is not surprising that little progress on identifying genetic factors underlying the common types of hyperlipidemia has been made.

These disorders may not necessarily occur as a result of an altered protein product, but rather as a result of subtle alterations in the level of expression of one or more of the many genes involved in lipid transport and metabolism. If the expression of the pathological phenotype occurs only in the presence of environmental factors or by the interaction of more than two genes, it is clear that it will be extremely difficult to identify the genetic components in the pathogenesis of the disease by simple pedigree analysis of the phenotype.

Recombinant DNA technology makes it possible to readily detect alterations of nucleotide sequence in the genome that are associated with pathological conditions. In some cases, these alterations occur within coding sequences and therefore result in an altered protein product, while in others, they are located in noncoding regions. The latter may be within regulatory sequences of the gene or, more likely be silent. In this case they may be physically linked to some other abnormality present in a neighbouring region of the chromosome, thus constituting useful genetic markers.

In the last decade, all the genes coding for key proteins involved in lipid transport and metabolism have been cloned and localized on human chromosomes[6]. These are 'candidate genes,' in other words genes at which allelic variations are likely to affect phenotype parameters. The availability of these genes not only give us the possibility of studying in depth the molecular pathology of dyslipidemias, as presented in other papers of this issue, but also provides material to develop innovative DNA based diagnostic tools to contribute to a better management of these diseases. In fact, knowing for example the genetic vulnerability of an individual it would be possible to commence early dietary intervention and avoid harsher therapeutic measures later in life.

We shall discuss new DNA based diagnostic tests aimed at analyzing either the genotype, in the form of polymorphic DNA markers, structural variants and regulatory mutants, or the phenotype, as for example plasma apolipoprotein levels determined by recombinant protein based tests.

GENOTYPE ANALYSIS BY DNA BASED DIAGNOSTIC TESTS

Genetic markers strategy

There are several basic procedures for detecting nucleotide sequence variations. The classical one is the restriction fragment length polymorphisms (RFLPs) approach through the Southern blotting technique.[7,8] This method is based on the sequence specificity of restriction endonucleases whose recognition sites are scattered throughout the chromosonal DNA. Their variations, inherited in a simple Mendelian fashion, can abolish a previous site or create a new one, thus providing a simple tool to analyze the genome, detecting rearrangements in the DNA sequence (addition, deletion, translocation, inversion) as well as substitution of a single nucleotide.

An alternative approach to detect nucleotide sequence variation is based on the use of synthetic oligonucleotides and differential hybridization conditions.[8] The specificity of this approach is based on the ability of short oligonucleotides to anneal, under the appropriate conditions, only to those sequences to which they are perfectly matched, a single base mismatch is sufficiently destabilizing to prevent hybridization. This technique, although a powerful and general method, requires microgram quantities of sample DNA, gel electrophoresis and highly radioactive probes due to the complexity of human genomic DNA.

These problems have been recently overcome by the development of the method of primer directed enzymatic amplification of specific genomic sequences (the polymerase chain reaction or PCR), which is capable of producing up to 10^6 copies of the target DNA in a single operation.[9,10] Repeated cycles of DNA denaturation, oligonucleotide primer annealing and extension by a thermostable DNA polymerase, result in an exponential accumulation of the specific target fragment. At this stage, the material can be subject to analytical procedures such as oligonucleotide hybridization, restriction enzyme mapping, DNA sequencing and mismatch specific recognition. The use of a thermostable DNA polymerase in the PCR technique, significantly improves the specificity and yield of the reaction, reduces costs and allows the development of an automated procedure.

Nucleotide polymorphisms, detected by any of these procedures, can be used as genetic markers to follow the inheritance of defective alleles in linkage analysis. Of course, there are cases where the polymorphisms are directly responsible for the disorder. The majority of these 'functional' polymorphisms in contrast to 'neutral' ones, will be intragenic, occurring in regions that affect either the transcriptional regulation of the gene or the quality of the final product.

An example of RFLP study in dyslipidemias: the apolipoproteins AI/CIII/AIV gene complex

The apolipoprotein AI gene was one of the first genes involved in the lipid transport and metabolism system to be cloned.[11,12] The Apo AI, CIII and AIV genes are clustered on chromosome 11.[13] The RFLPs described in this gene complex are shown in Figure 1 and have been used to look for associations with CVD and different forms of dyslipidemias in several populations. The results obtained by different groups have been variable and often contrasting, probably due to differences in accurately defining the phenotypes and to the genetic heterogeneity of the populations studied.

We shall discuss in detail the example of the SstI polymorphism and its associations with hyperlipidemia. The Sst I RFLP distinguishes between two alleles of this gene complex in the population: S1 or Apo AI/4.2 and S2 or Apo AI/3.2.[5] On restriction with endonuclease SstI, the S1 allele yields two fragments of 5.7 and 4.2 kb, which hybridize to an Apo AI cDNA probe, whereas the S2 allele yields fragments of 5.7 and 3.2 kb. The S2 allele is defined

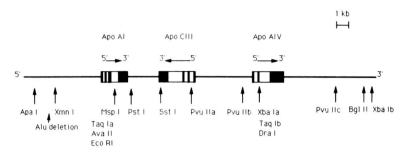

Fig. 1 Apolipoprotein AI/CIII/AIV gene complex. Locations of restriction enzyme sites are indicated by arrows: if position is not known precisely, sites are placed under probe sequences. Figure modified from Reference 6.

by a substitution of a cytosine residue for a guanosine at the 40th position of the 3' noncoding region of the Apo CIII gene,[17] resulting in a new Sst I site. This indicates that this change may be a neutral polymorphism, representing only a genetic marker.

The frequency of the two Apo AI alleles was originally estimated for both normolipidemic and type IV/V hypertriglyceridemic patients of Caucasian origin.[15] In this preliminary study the S2 allele was not found in normolipidemic individuals but only in hypertriglyceridemic subjects. However, in other racial groups, a large difference is seen in the distribution of this allele in normolipidemic individuals. For example, 65% of Chinese and 35% of Japanese normolipidemic individuals possess this allele.[16] These data stress that caution is needed using RFLPs in association studies in heterogeneous population samples.

On the other hand, an extensive and careful study has recently confirmed the validity of the association of the S2 allele to hyperlipidemia in a Caucasian population,[17] as shown in Table 1. This work indicates that 33.3% of English patients with primary hyperlipidemia (excluding FH) had the S2 allele, compared to 6.1% of normolipidemic individuals. The increased frequency of this allele was statistically significant in each of the hyperlipidemic groups examined: type IIA (excluding patients with FH) type IIB and type IV.

The S2 RFLP seems to be associated with another yet undefined functional mutation present in a neighbouring region of the chromosome, predisposing to certain forms of hyperlipidemias. The association of a DNA polymorphism with any given genetic disease can open many perspectives for future clinical research. It should be possible, for example, to examine whether there is any significant

Table 1 Frequency of the S2 genotype in a Caucasian population sample.[16] Groups are divided according to Fredrickson classification, differentiating within the IIA patients those individuals with classical familial hypercholesterolemi (FH). AI/CIII genotype, number of individuals per group (N) and S2 carriers frequency are indicated

Group	AI/CIII genotype	N	% of group with S2
Normolipidemic	S1/S1	46	6.1
	S1/S2	3	
IIA with tendon xanthomas (FH)	S1/S1	36	14.3
	S1/S2	6	
IIA without tendon xanthomas	S1/S1	23	39.5
	S1/S2	15	
IIB	S1/S1	60	31.0
	S1/S2	27	
IV	S1/S1	21	32.3
	S1/S2	10	
Random controls	S1/S1	50	7.4
	S1/S2	4	

difference in either the clearance rate of dietary triglyceride from the circulation, or in the severity and incidence of cardiovascular disease among dyslipidemic patients with different genotypes. A single genetic marker may not be enough to define a aetiological genotype, but the extensive use of different genetic markers located on the same chromosome (haplotypes), can produce detailed genotyping of individuals, as it was originally done in beta thalassemia.[18] In the case of polygenic diseases such as atherosclerosis or dyslipidemias, the combination of haplotypes at different loci might provide a comprehensive evaluation of the risk.

Identification and analysis of polymorphisms in regulatory elements controlling gene expression

The characterization of regulatory elements controlling at the molecular level the expression of genes coding for proteins involved in lipid metaboiism and transport is an alternative approach to polymorphisms analysis. The identification of these sequences give us the opportunity to look for variations that may be present in affected individuals, defining genetic markers with a functional significance for molecular diagnosis.

The presence of a mature and active protein is the result of a complicated pathway that is usually summarized by the term 'gene expression'. A general scheme of the flow of genetic information is shown in Figure 2. The entire information for these steps is encoded in the DNA sequence and the production of mature mRNA represents the main method by which gene expression is regulated.[19] As can be seen in Figure 2, there are many steps where this process can be controlled. The DNA sequences (*cis*-acting elements) involved in the initiation of gene transcription can be classified into those which confer basic transcriptional competence and those involved in the spatial and temporal regulation of that competence. Given an 'open' chromatin configuration in which the gene is accessible to the transcription machinery, the basic level of transcriptional activity of a gene is determined by an element called the promoter. It is by means of this element that

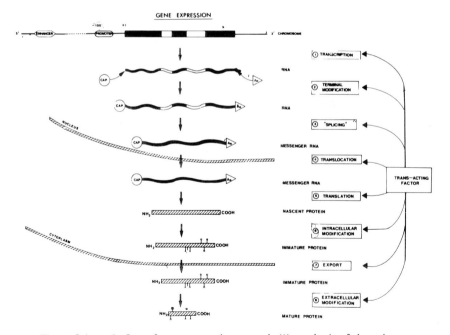

Fig. 2 Schematic flow of gene expression control: (1) synthesis of the primary transcripts; (2–3) turnover and processing of the primary transcript; (4) transport of the mRNA out of the nucleus and turnover in the cytoplasm; (5) translation of the mRNA into protein; (6–8) modification and turnover of the protein.

the transcription initiation site (CAP site) and direction of a particular gene are recognized by the RNA polymerase complex. The promoter is located within about 100 bp upstream of the site of transcription initiation and its activity is dependent upon position and orientation. The 'classic' core promoter is composed of two conserved sequence elements, the 'TATA' and 'CAAT' boxes. However, in some cases these motifs are associated or even replaced by GC rich repeat structures. Enhancer or activator elements are defined as *cis*-acting DNA sequences which dramatically stimulate transcription from homologous and heterologous promoters in an orientation and, within limits, distance independent manner. Although different enhancers share little sequence homology, a limited number of short, degenerate and often repeated consensus sequences have been identified, suggesting that enhancers represent mosaics of these possibly evolutionary related sequence motifs: a modular organization which would enormously increase the combinational possibilities of transcriptional control of gene expression. However, such consensus sequences occur in regions of DNA which do not have enhancer properties and thus at present enhancers can only be functionally defined.

Several cellular enhancers can activate transcription either in a tissue-specific manner or in response to an inducing signal. Transcriptional enhancement involves the interaction of enhancers with specific trans-acting factor(s). They appear to be required both to drive the transcription of genes encoding major products of terminally differentiated cells and to direct the strong transient expression of certain genes in response to a specific extracellular stimulus or at a particular stage of development. From the flow of gene expression, summarized in Figure 2, it is obvious that sequence mutations in any of the control elements of these single steps will have an effect on gene expression that could range from its complete abolishment to subtle variations on the level and/or timing of expression. To study variations in these sequences that may result in pathological phenotypes, it is necessary first to accurately map and sequence the critical regions, particularly focusing on promoters and enhancers, and then look for sequence variations in normal and pathological populations.

Putative regulatory mutant in the apolipoprotein AI gene

As previously mentioned, a number of studies have shown RFLPs in the Apo AI/CIII/AIV gene complex that associate in patient groups with dyslipidemia and/or atherosclerosis.[6] In particular, several reports suggest that polymorphisms at this locus are associated with low high density lipoprotein cholesterol (HDL-C) levels and coronary heart disease.[20-24]

The main protein components of HDL are apolipoproteins AI and AII. We have focused our attention on the 5′ flanking region of the Apo AI gene where we have identified an apolipoprotein AI gene promoter polymorphism due to an adenine to guanine transition, located in position −78 from the transcription start site 51 bp upstream from the 'TATA' box, as shown in Figure 3A.[17,25]

Extending this analysis to large populations was difficult because of the lack of convenient RFLPs. However, the PCR amplification technique makes possible to study sequence variations in a large

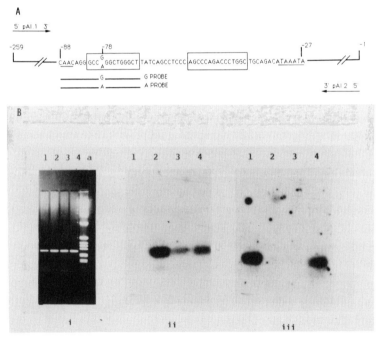

Fig. 3 (**A**) The 5′ flanking region of the AI Apo gene. Numbers indicate position of nucleotides from transcription start site. The presumed 'CAAT' and 'TATA' boxes are underlined. The A/G variation is shown in position −78 along with the two oligonucleotides, probes A and G, used in allele specific hybridization. The 14/15 bp inverted repeat sequences are boxed. (**B**) Genotype analysis of amplified DNA by allele specific oligonucleotide hybridization: (i) agarose gel, stained with ethidium bromide; (ii and iii) Southern blots probed with allele specific oligonucleotides probe G and A, respectively. Lane 1–4: amplified DNA preparation with different genotypes (1 = A/A, 2 = G/G, 3 = G/G, 4 = A/G). Lane (a) MW size markers.

number of subjects. Amplification of the promoter region of Apo AI gene followed by specific-oligonucleotide hybridization, as shown in Figure 3B, allowed us to analyze the frequency of the alleles previously described, and for convenience named A and G alleles, according to the nucleotide present in that position. The distribution of the polymorphism was studied in three different groups of subjects from each sex selected from an epidemiological survey on atherosclerosis risk factors, and divided according to their HDL-C concentration, as shown in Table 2.[26]

In women the group with high HDL-C levels showed an allelic frequency for the A polymorphism (0.27), significantly higher when compared to the intermediate HDL-C concentraton group (0.14) and the lower HDL-C concentration group (0.10). In men the distribution of allelic frequencies of the A/G polymorphism was not affected by HDL-C levels. No statistically significant difference was observed in the distribution of allelic frequencies of the A and G polymorphism between men and women, as a whole. The genotype distribution is in Hardy-Weinberger equilibrium for both sexes.

Thus the different distribution of allelic frequencies as a function of HDL-C concentration between the two groups is not due to a selection effect, and other genetic or environmental factors should account for this difference.

Table 2   Genotype distribution and relative allelic frequencies for the A and G alleles in different groups stratified by HDL cholesterol (HDL-C) levels

Percentile of HDL-C	Genotype				Allelic frequencies	
	n	GG	AG	AA	G	A
WOMEN						
< 10th	40	32	8	—	0.90	0.10
10th–90th	45	33	11	1	0.86	0.14
> 90th	51	27	20	4	0.73	0.27*
Total	136	92	39	5	0.82	0.18
MEN						
< 10th	42	36	5	1	0.92	0.08
10th–90th	32	21	11	—	0.83	0.17
> 90th	34	26	7	1	0.86	0.14
Total	108	83	23	2	0.87	0.13

n = numbers of subjects
* highest decile v. lowest decile, chi-square = 7.54 P < 0.006;
highest decile v. 10th–90th percentile, chi-square = 4.07 P < 0.04

The association of the A allele with high HDL-C and hence Apo AI levels, in women but not in men, may be due to the fact that the polymorphic site is in a sequence of the Apo AI gene promoter somehow involved in the response to the cyclic hormonal stimulation characteristic of premenopausal women. The DNA region surrounding the polymorphism is a 51 bp fragment that is G-C rich and contains an inverted repeat composed of two 14/15 bp elements. The homology and self complementarity of the inverted repeats is disrupted when G is present instead of A in position −78. Both direct and inverted repeat sequences with peculiar base composition within 5′ flanking regions have been implicated in the regulation of expression of various genes.[27–29] Site specific mutagenesis within these repeated elements either naturally occurring, as in the human beta globin gene,[30] or produced by in vitro mutation, as in the mouse beta globin gene,[31] have been shown to reduce transcriptional activity. In order to clearly define the relationship between this promoter polymorphism, and HDL-C and Apo AI levels, we are currently performing in vitro studies on the expression of the two different alleles and following their segregation in kindreds of hyperalphalipoproteinemic subjects.

Identification of apolipoprotein AII regulatory sequences

Whilst there are several examples of promoter mutations in molecular pathology, at present there is no example of enhancer mutations. This is obviously due to the recent definition of these elements and the scant information on their variants in the population. However an approach similar to the one described above can be applied to the identification of enhancers and tissue specific determinants, and analysis of their mutants.

The apolipoprotein AII gene provides an interesting example of this kind of analysis. It is transcribed exclusively in the adult liver and intestine, and the resulting transcripts appear to be identical. Apo AII is the only apolipoprotein not expressed in the gut of 6 to 12 week human embryos, indicating that the expression of this gene is also subject to specific developmental regulation.[32] Studies of transient expression of deletion mutants in the 5′ flanking region of the Apo AII gene showed that this dual tissue-specific expression is regulated by an enhancer element of 201 bp. This sequence is situated between positions −853 to −653 upstream from the transcription initiation site, its presence being absolutely required for transcription from the Apo-AII promoter.[29]

Enhancer elements are composed of a modular set of DNA sequence motifs that usually constitute binding sites for nuclear proteins. A general method of analyzing these elements is to study their interaction with specific transcription factors present in the cell nucleus. Obviously, each cell type has a set of specific transcription factors that either stimulate or inhibit the expression of particular genes. In the case of apolipoprotein AII gene enhancer, the identification of the binding sites for the hepatocyte nuclear proteins that control the cell specific transcription has recently been described.[33] Five different binding sites (motifs I to V) are present and specifically interact with potential *trans*-acting proteins (Fig. 4). In order to differentiate between cell type specific and general *trans*-acting factors, the binding studies were simultaneously carried out with liver (specific) and HeLa (non specific) nuclear extracts.

Several proteins from liver or HeLa nuclear extracts have been shown to bind to different DNA sequences present in this tissue specific enhancer (Fig. 4). The motifs III, IV and V can be occupied differently by liver or HeLa nuclear proteins and two hypersensitive zones generated by protein-DNA interactions (between motifs II-III and IV–V) are present only when liver nuclear extracts were tested. These different binding sites share specific DNA sequences with 5′ regulatory regions of other apolipoprotein genes, thus indicating that their expression might involve common control mechanisms.

The identification of these essential elements will allow us to screen for their variants in normal and dyslipidemic populations, as they should affect the specific apolipoprotein synthesis and hence its plasma levels. There are two alternative technical approaches for this type of study, both starting from direct PCR amplification of the segment. The first one is labour intensive and consists of sequencing each allele, the second approach is to look for sequence variations using for example a recently developed chemical degradation technique.[34] In this latter strategy the amplified DNA is hybridized with a wild-type DNA sequence and a specific chemical degradation for mismatched bases is carried out, using hydroxylamine (for C mismatch) or osmium tetroxide (for C and T mismatches). This technique is very precise and allows a fast analysis of sequence variants in a large number of subjects.

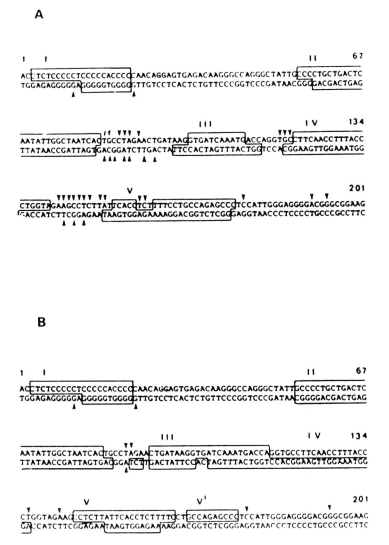

Fig. 4 Localization of the protected regions in the DNA sequence of the human Apo AII enhancer. In (**A**) using human or rat liver nuclear protein extract; in (**B**) using HeLa nuclear protein extracts. Numbers indicate the enhancer nucleotides positions. The protected sequences, named by roman numbers, are boxed (motifs). Arrow heads indicate sizes of enhanced DNAse I cleavage.

PHENOTYPE ANALYSIS BY DNA BASED DIAGNOSTIC TEST

Genetic analysis will eventually lead to the identification of the gene variants involved in the predisposition to dyslipidemias and cardiovascular diseases. The expectation is to establish a clear relationship between individual alleles and biochemical and clinical phenotype. To the most standardized biochemical measurements, as total cholesterol and triglycerides plasma levels, have been added in the last decade more predictive parameters for CVD risk, such as HDL-cholesterol and LDL-cholesterol levels. These data revealed the great importance of having well defined biochemical parameters as preliminary and necessary tools to investigate further the genetic traits underlying the disease. Furthermore there is a strong indication that the apolipoprotein AI/B ratio represents an even more powerful index to assess risk of cardiovascular disease.[35,36] In fact, from a clinical point of view, apolipoprotein determinations are an essential parameter for the diagnosis of disorders in lipid metabolism and for providing a more precise evaluation of the individual coronary risk. This increased demand for apolipoprotein measurement requires a technique suitable for routine use in clinical laboratories. Several methods are currently used for these analyses, including radial immuno diffusion (RID), immuno nephelometry (INA), radio immunoassay (RIA) or enzyme-linked immunoassay (ELISA).[37]

Among these techniques, the immunoenzymatic assays have several advantages, due to their high sensitivity, specificity, flexibility as well as not requiring radioactive tracers. In a current ELISA assay highly purified antigen and marker enzyme (generally conjugated to an antibody) are needed and separately used. Moreover the chemical coupling of the antibody and the tracer gives heterogeneous mixtures as conjugation takes place at various points.

An attractive alternative to the chemical coupling is the preparation of antigen-enzyme hybrid proteins by recombinant DNA techniques. In fact, the expression of chimeric genes in *E. coli* allows the production and purification of these proteins in large amount, with all the required biochemical and immunological characteristics.[38,39]

A practical application of the use of this technology has been recently developed.[40] The human apolipoproteins AI or B moieties were fused to an enzymatically active beta-galactosidase by genetic engineering and the chimeric proteins expressed in *E. coli*.

The purified fusion proteins, characterized by high enzymatic activity and apolipoprotein antigenicity, were used to set up a recombinant immunoenzymatic competition assay (RIECA) for the quantitative determination of apolipoprotein AI and B, as shown in Figure 5.

This assay is based on a microtiter wells plate coated with specifically obtained monoclonal antibodies against apolipoproteins AI and B. The principle of the test consists in an immunocompetition reaction, for the monoclonal antibody, between the natural apolipoprotein in the serum sample and the correspondent recombinant apolipoprotein linked with the enzymatic activity. In

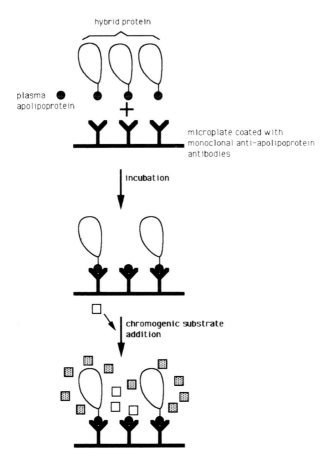

Fig. 5 A general scheme for the recombinant immuno enzymatic competition assay (RIECA) in the quantitative determination of apolipoprotein AI and B.

these conditions the quantitative determination of Apo AI and B levels is carried out by a colorimetric reaction given by the beta-galactosidase activity. In particular, the hybrid tracer proteins are employed in saturating amounts with respect to the monoclonal antibody, so that the enzymatic activity measured by the spectrophotometer is negatively correlated with the apolipoproteins concentration of the sera samples.

The construction of plasmids coding active β-galactosidase fused with aminoacid sequences of human apolipoproteins AI and B is indicted in Figure 6. The hybrid proteins ISMAI and ISMB1, that aggregate in vivo to form intracellular insoluble 'inclusion bodies', were purified by a denaturtion/renaturation process. Through this purification process it was possible to obtain about 1 g of purified protein per 10 l fermentation mixture in which ISMAI and ISMB1 proteins represent about 80% and 60% of the final product, respectively.

Calibration curves for the assay were obtained by using increasing amounts of apolipoproteins standard serum and a constant concentration of hybrid proteins. In a semi-logarithmic plot, in which the serum apolipoprotein concentrations were expressed as $\log_2$ concentration, the range of linearity for Apo I and Apo B determinations was between 31.2–250 mg/dl and 18–150 mg/dl, respectively, amply covering the concentration range found in clinical practice. Using serial dilutions of standards, the sensitivity of the assay was calculated to be 0.1 mg/dl.

In order to verify the validity of this new method, a comparison with radial immuno diffusion (RID) method was performed. A good correlation (R) was obtained: 0.91 and 0.89 for Apo AI and Apo B determinations, respectively. In another study the concentrations of Apo AI and Apo B were found to correlate with HDL (R = 0.73) and LDL (R = 0.93) plasma levels, respectively. These results indicate that apolipoprotein values determined with RIECA satisfactorily correlate with their corresponding lipoprotein classes.

The principle of this DNA based assay is very flexible and allows one to design and produce diagnostic systems of potential interest for any protein whose gene is cloned and specific monoclonal antibodies are available. Furthermore this assay might provide a more precise and informative tool, giving the opportunity not only to measure total protein levels, but also to discriminate among mutant forms of the same protein, taking advantage of the specificity of the monoclonal antibody and the tracer protein.

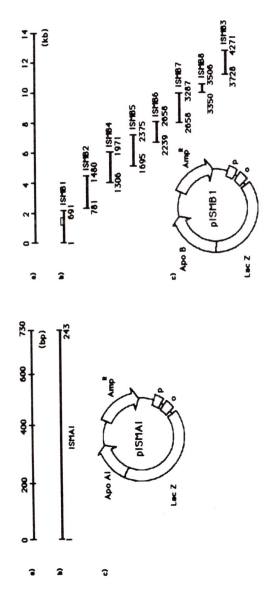

Fig. 6 Representation of the pISMAI and pISMB1 expression vectors and their construction from Apo AI and Apo B coding sequences, respectively; (**a**) DNA sequence coding for mature apolipoprotein; (**b**) amino acid sequence of the insert (**aa 1** as the first **aa** of the mature protein); (**c**) expression vector, coding active beta-galactosidase/apolipoproteins hybrid proteins.

ACKNOWLEDGEMENTS

We are very grateful to our collegues C C Shoulders, M Lucero, F Pagani, C S Shelley, C Tosi, I Giuntini and C Mariani who extensively contributed to the research work reviewed in this paper.

REFERENCES

1 White R, Caskey CT. The human as an experimental system in molecular genetics. Science 1988; 240: 1483–1488
2 Landegren U, Kaiser R, Caskey CT, Hood L. DNA diagnostics: molecular techniques and automation. Science 1988; 242: 229–237
3 Antonarakis SE. Diagnosis of genetic disorders at the DNA level. N Engl J Med 1989; 320: 153–163
4 Breslow JL. Genetic basis of lipoprotein disorders. J Clin Invest 1989; 84: 373–380
5 Shoulders CC, Baralle FE. Genetic polymorphism in the Apo AI/CIII complex. Methods Enzymol 1986; 128: 727–745
6 Lusis AJ. Genetics factors affecting blood lipoproteins: the candidate gene approach. J Lipid Res 1988; 29: 397–429
7 Southern EM. Detection of specific sequences among DNA fragments separated by gel electrophoresis. J Mol Biol 1975; 98: 503–517
8 Weatherall DJ. The new genetics and clinical practice. Oxford Univ Press, 1985
9 Erlich HA edt. PCR technology: principles and applications for DNA amplification. Mac Millan, 1989
10 Eisestein BI. The polymerase chain reaction: a new method of using molecular genetics for medical diagnosis. N Engl J Med 1990; 322: 178–181
11 Breslow JL, Ross D, McPherson J et al. Isolation and characterization of cDNA for human apolipoprotein AI. Proc Natl Acad Sci USA 1982; 79: 6861–6865
12 Shoulders CC, Baralle FE. Isolation of the human HDL apolipoprotein AI gene. Nucleic Acids Res 1982; 10: 4873–4882
13 Karathanasis SK. Apolipoprotein multigene family: tandem organization of human apolipoprotein AI, CIII and AIV genes. Proc Natl Acad Sci USA 1985; 82: 6374–6378
14 Sharpe CR. Ph.D. Thesis. Oxford University, 1985
15 Rees A, Shoulders CC, Stocks J, Galton DJ, Baralle FE. DNA polymorphism adjacent to human apoprotein AI gene: relation to hypertriglyceridaemia. Lancet 1983; i: 444–446
16 Rees A, Stock J, Sharpe CR et al. Deoxyribonucleic acid polimorphism in the apolipoprotein AI-CIII gene cluster: association with hypertriglyceridaemia. J Clin Invest 1985; 76: 1090–1095
17 Shoulders CC, Ball MJ, Baralle FE. Variation in the apo AI/CIII/AIV gene complex its association with hyperlipidemia. Atherosclerosis 1990
18 Orkin SH, Kazazian HH, Antonarakis SE et al. Linkage of β-thalassaemia mutations and β-globin gene polymorphism with DNA polymorphism in human β-globin gene cluster. Nature 1982; 296: 627–631
19 Lewin B. Genes. Chichester: Wiley, 1983
20 Ordovas JM, Schaefer EJ, Salem D et al. Apolipoprotein AI gene polymorphism associated with premature coronary artery disease and familial hypoalphalipoproteinemia. N Engl J Med 1986; 314: 671–677
21 Sidoli A, Giudici G, Soria M, Vergani C. Restriction fragment length polymorphism in the apolipoprotein AI-CIII gene complex occurring in a family with hypoalphalipoproteinemia. Atherosclerosis 1986; 62: 81–87
22 Anderson RA, Benda TJ, Wallace RB, Eliason SL, Lee J, Barnes TL. Preva-

lence and association of apolipoprotein AI-linked DNA polymorphisms: results from a population study. Genet Epidemiol 1986; 3: 385–397

23 Deeb S, Failor A, Brown BG, Brunzell JD, Albers JJ, Motulsky AG. Molecular genetics of apolipoproteins and coronary heart disease. In: Cold Spring Harbor Symp Quant Biol 1986; LI: 403–409

24 Price WH, Kitchin AH, Morris SW, Burgon PRS, Wenham PR, Donald PM. DNA restriction fragment length polymorphisms as markers of familial coronary heart disease. Lancet 1989; i: 1407–1410

25 Sharpe CR, Sidoli A, Shelley CS, Lucero MA, Shoulders CC, Baralle FE. Human apolipoproteins AI, AII, CII and CIII, cDNA sequences and mRNA abundance. Nucleic Acids Res 1984; 12: 3917–3932

26 Pagani F, Sidoli A, Lucero MA, Zakin MM et al. Human apolipoprotein AI gene promoter polymorphism: association with hyperalphalipoproteinemia. J Lipid Res 1990 in press

27 Maniatis T, Goodbourn S, Fisher JA. Regulation of inducible and tissue specific gene expression. Science 1987; 236: 1237–1245

28 McKnight SL, Kingsbury R. Transcriptional control signals of a eukaryotic protein coding gene. Science 1982; 217: 316–324

29 Shelley CS, Baralle FE. Dual tissue specific expression of apo AII is directed by an upstream enhancer. Nucleic Acids Res 1987; 15: 3801–3821

30 Wong C, Antonarakis SE, Goff SC, Orkin SH, Boehm CD, Kanazazian HH. On the origin and spread of beta thalassemia: recurrent observation of four mutations in different ethnic groups. Proc Natl Acad Sci USA 1986; 83: 6529–6532

31 Myers MA, Tilly T. Fine structure genetic analysis of a beta globin promoter. Science 1986; 232: 613–618

32 Hopkins B, Sharpe CR, Baralle FE, Graham CF. Organ distribution of apolipoprotein gene transcripts in 6–12 week postfertilization human embryos. J Embryol Exp Morph 1986; 97: 177–187

33 Lucero MA, Sanchez D, Ochoa AR et al. Interaction of DNA-binding proteins with the tissue-specific human apolipoprotein-AII enhancer. Nucleic Acids Res 1989; 17: 2283–2299

34 Cotton RGH, Rodriguez NR, Campbell RD. Reactivity of cytosine and thymine in single-base-pair mismatches with hydroxilamine and osmium hydroxide and its application to the study of mutations. Proc Natl Acad Sci USA 1988; 85: 4397–4401

35 Avogaro P, Bittolo Bon G, Cazzolato G, Quinci GB. Are apolipoproteins better discriminators than lipids for atherosclerosis? Lancet 1979; i: 901–903

36 Van Stiphout WA, Hofman A, Kruijssen HA, Vermeeren R, Groot PH. Is the ratio Apo B/Apo AI an early predictor of coronary atherosclerosis? Atherosclerosis 1986; 62: 179–182

37 Bachorik PS, Kwiterovich PO. Apolipoprotein measurements in clinical biochemistry and their utility vis-a-vis conventional assays. Clin Chim Acta 1988; 178: 1–34

38 Ullmann A. One step purification of hybrid proteins which have β-galactosidase activity. Gene 1984; 29; 27–31

39 Bulow L. Characterization of an artificial bifunctional enzyme, β-galactosidase/galactokinase, prepared by gene fusion. Eur J Biochem 1987; 163: 443–448

40 Galliani S, Tosi C, Giuntini I et al. RIECA: recombinant immuno enzymatic competition assay for apolipoproteins AI and B. Anal Biochem 1990 in press

British Medical Bulletin (1990) Vol. 46, No. 4, pp. 960–985
© The British Council 1990

Pathology of atherosclerosis

N Woolf

Department of Histopathology, University College and Middlesex School of Medicine, University College London, London, UK

This communication gives a brief account of the morphology and natural history of atherosclerosis. It defines atherosclerosis, in dynamic terms as the resultant of three interacting sets of processes: accumulation and modification of plasma-derived lipid within the arterial intima, connective tissue proliferation and connective tissue necrosis forming an atheromatous pool at the plaque base. The first of these leads to the accumulation of lipid-filled macrophages within the affected intima and this step is most probably mediated via oxidative modifications of the low density lipoprotein molecule. An alteration of the functional phenotype of the intimal smooth muscle cell as a result of interactions with growth factors (most notably PDGF) constitutes the basis for the connective tissue proliferation. Plaque necrosis, which is extremely important as a risk factor for acute thrombosis, is the least well understood area; the activated macrophage may well play a significant role in this connexion.

DEFINITION OF ATHEROSCLEROSIS IN TERMS OF MORPHOLOGY AND PROCESS

In morphological terms, atherosclerosis may be defined as 'a widely prevalent arterial lesion characterized by patchy thickening of the intima, the thickenings comprising accumulations of fat and layers of collagen-like fibres, both being present in widely varying proportions'.[1]

This definition emphasizes certain important points:

1. The focal distribution of lesions which is almost certainly governed by haemodynamic factors

0007–1420/90/0046–0960/$10.00

2. The predominant, though not exclusive involvement of the tunica intima rather than the deeper layers of the arterial wall

3. The complex nature of atherosclerotic plaques consisting partly of a lipid-rich pool of necrotic connective tissue at the plaque base and partly of a fibro-muscular 'cap' on the luminal aspect of the atheromatous pool

Interpretation in terms of biological mechanisms rather than in those of morphology might lead one to redefine atherosclerosis as the result of the following processes.

(a) Excess infiltration and/or retention of plasma-derived lipid within the arterial intima and the subsequent modification of this lipid

(b) Proliferation of connective tissue resulting in the formation of a sub-endothelial fibro-muscular cap which may compromise the integrity of the arterial lumen to a considerable degree

(c) Necrosis of connective tissue at the plaque base leading to the formation of a soft, deformable atheromatous pool. If this is of massive proportions in relation to the fibro-muscular cap, plaque rupture may occur.

Intimal fibro-muscular proliferation is the mediator of arterial stenosis such as is seen in the coronary arteries of patients with stable angina. Plaque necrosis, if severe and extensive, is a potent risk factor for serious or life threatening clinical situations such as unstable angina, myocardial infarction or the onset of serious ventricular arrhythmias. These are related to the rapid intra-mural and intra-luminal thrombosis resulting in most instances from rupture or erosion of the plaque cap.[2-4]

MORPHOLOGY OF ATHEROSCLEROTIC PLAQUES

The earliest events in human atherogenesis are largely a matter of guesswork. There is still controversy as to what constitutes the earliest lesion of human atherosclerosis, no certainty that the common endpoint of a raised fibrolipid plaque may not be reached from different beginnings and no agreement that the fatty lesions, which many regard as precursors of atherosclerotic plaques, inevitably develop into the latter.[5]

In a number of animal models of atherogenesis, one of the earliest structural changes is adhesion of monocytes to the endothelial surface this being followed by entry into the subendothelial tissues where they can ingest large amounts of lipid and become converted to foam cells.[6-9] In one model, the White Carneau pigeon, which shows a marked susceptibility to develop atherosclerosis, independent of diet, monocyte adhesion is associated with increased turnover of endothelial cells.[10] Though the sequence of events in human lesion development is not known there is good morphological evidence that here too monocyte traffic between the blood and the artery wall occurs.

The fatty streak

Fatty streaks, probably the earliest macroscopically recognizable lesions, are found from childhood onwards. In contrast with mature fibrolipid plaques, there is no significant variation in prevalence and severity in different populations and this discordance presents a challenging problem in interpretation. Measurement of the extent of intimal surface involvement and the distribution pattern of fatty streaks is facilitated if the vessel to be studied is stained with a fat-soluble dye. The lesions appear to start as small (1–2 mm in diameter) rounded or oval, yellowish dots minimally elevated above the surface of the adjacent intima. These dots tend to occur in rows roughly parallel to the streamlines of the flowing blood and coalesce to form streaks along the long axes of the affected artery. Fatty streaks in the aorta can be seen in more than 40% of infants coming to necropsy between the ages of 1 and 12 months.[11] In the youngest children the lesions are localized to the region of the aortic valve ring, the area of the ductus scar and the ostia of the intercostal vessels. As the child grows, the aortic arch and the posterior wall of the thoracic aorta become more severely affected, following which streak lesions appear in the abdominal aorta. In the coronary artery tree definite fatty streaks appear in the proximal segments of the vessels at about the time of puberty.

Distribution pattern of fatty streaks

In the young, fatty streaks are concentrated in the arch and upper part of the ascending aorta in a rough fan shape. From a point about 5–7 cm distal to its most proximal portion, the 'fan' narrows and the streaks become most obvious on the posterior wall of the aorta. Distal to the main visceral aortic branches, the distribution pattern changes and the whole circumference of the aorta becomes involved in an apparently haphazard way. In the thoracic aorta of young humans, the streak lesions are most apparent in the proximal portions of the intercostal vessel ostia and there is relative sparing of the distal edges or 'flow-divider' regions of the ostia. A similar pattern in relation to intercostal arteries is seen in hyperlipidaemic non-human species provided only that the cholesterol concentrations in blood and tissues are in a steady state.[12-14] Thus there appears to be a predilection for lipid deposition and streak lesion formation to occur in those parts of the branch where the wall shear rates (the velocity gradients between layers of fluid in contact with and near, the artery wall) are low and where non-atherosclerotic intimal thickening is much greater than in the high wall shear rate flow divider.

Histological examination of the fatty streak shows localized thickening of the intima this being associated with the presence of fat droplets seen most easily in frozen sections stained with fat soluble dyes. The localization of the stainable fat appears to be predominantly intra-cellular though, in some lesions, there is a 'dusting' of sudanophilic material along the course of the internal elastic lamina marking the intima/media boundary.

Cell population of the streak lesion

The streak lesion is more cellular than the non-involved intima though it is not possible to determine the nature of the cells on light microscopy unless appropriate immunohistochemical methods are used. When this is done a significant proportion of the cells is seen to originate from monocytes and to show macrophage markers.[15-18] These results find morphological support from electron microscopic examination of fatty streaks from which at least two cell types can be identified. One of these is quite clearly an intimal smooth muscle cell which can be recognized by its elongated cell profile, by the presence of 'dense bodies' (the analogue of Z bands) situated at the periphery of the cell and from evidence of basement membrane formation. The second cell type lacks an elongated profile and has a ruffled and convoluted plasma membrane. The cytoplasm contains numerous lipid vacuoles and occasional 'myelin figures' are seen, all these features suggesting that these cells are macrophages (Fig. 1). Despite these observations, it must be remembered that once a cell becomes heavily loaded with lipid, recognition based on morphology may become impossible and in this situation one must rely on immunohistochemical identification of appropriate differen-

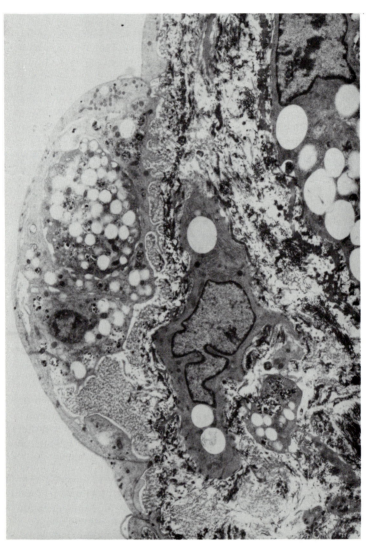

Fig. 1 Transmission electron micrograph of a small fatty streak in a congenitally hyperlipidaemic rabbit of the St Thomas's Hospital strain. The endothelium is deformed from beneath by the presence of lipid-laden macrophages. Deep to these is an intimal smooth muscle cell showing very obvious peripheral dense bodies. This cell, too, shows some evidence of lipid loading in the form of large vacuoles. This photograph illustrates the double nature of the foam cell population of fatty streaks ($\times 4,000$).

tiation markers. The recognition that the macrophage forms a significant element in the cell population of fatty streaks and, indeed, of mature atherosclerotic lesions also, is an important advance towards a fuller understanding of mechanisms involved in atherogenesis since the macrophage may exercise a range of functions within the arterial intima which far transcend its conventional role as a phagocytic cell.

Just as transmission electron microscopy and immunochemistry have increased our knowledge of the cell population of the fatty streak, so can scanning electron microscopy increase our appreciation of events taking place at the blood/vessel wall interface during lesion formation and progression.

For instance, in young hyperlipidaemic rabbits, scanning electron microscopy shows localized areas of subendothelial swelling whicha are particularly prominent in the intima situated just upstream of the intercostal artery ostia where, as already stated, wall shear rates are low. Somewhat longer periods of hyperlipidaemia are associated with the appearance of focal defects in the integrity of the endothelial lining and these defects are associated with the presence of large cells with markedly ruffled plasma membranes, the appearances of which suggest that they are macrophages (Fig. 2). Not infrequently, individual macrophages can be seen to penetrate between individual endothelial cells and in some areas, groups of such macrophages can be seen protruding into the arterial lumen from under 'bridges' of intact endothelial cells.[9] When human coronary arteries are examined by scanning electron microscopy fatty dots which are the precursors of the streak lesion, can be seen as small plateau-like subendothelial elevations some of which are associated with adherent leucocytes.[19]

Fate of the fatty streak

Controversy still clouds the issue as to whether fatty streaks progress to mature fibrolipid atherosclerotic lesions or not.[20] Clearly not all fatty streaks can behave in this way since the extent of intimal surface involvement by fatty streaking seen at necropsy in young subjects shows no significant differences between populations in which there is a high prevalence of severe atherosclerosis and those in which extensive and severe atherosclerosis is not seen. Careful histological examination of human lesions shows intermediate type lesions which have been interpreted as suggesting that progression from fatty streak to fibrolipid lesion can occur

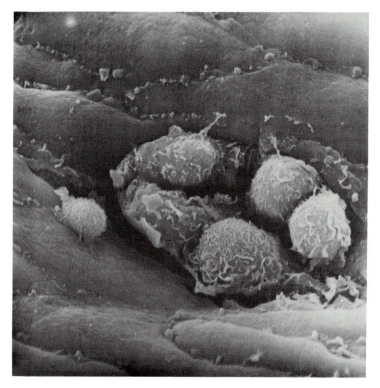

Fig. 2 Scanning electron micrograph of the intimal surface of the aorta in a hyperlipidaemic rabbit. The centre of the photograph is occupied by a well-demarcated defect in which five macrophages showing ruffled plasma membrane can be seen (× 2,400).

and a recent study[7] on longterm effects of diet-induced hyperlipidaemia on the arteries of non-human primates provides persuasive evidence of the correctness of this view. This combination of morphological and epidemiological data suggests that some fatty streaks originating early in life undergo regression while others progress to mature atherosclerotic plaques. The latter course of events is, presumably, more likely to occur in those populations whose degree of exposure to 'risk factors for atherogenesis' such as hyperlipidaemia, cigarette smoking, high blood pressure and diabetes mellitus is great.

At the fatty streak stage it is very difficult, if not impossible, to distinguish between those lesions likely to progress and those which will not. In a histological study carried out on fatty streaks derived from young deceased subjects from several different populations, in which the discordance between the degree of fatty streaking in the young and the degree of involvement by fibrolipid plaques in middle-aged and elderly subjects was greatest,[21] the presence of focal necrosis and of a severe inflammatory cell infiltrate within the lesion was equated with the likelihood of progression.

Raised lesion or fibrolipid plaque

The raised lesion or fibrolipid plaque is the archetypal lesion of atherosclerosis and complications of this lesion, most notably plaque fissure or ulceration, constitute the basis for the vast majority of cases of occlusive arterial disease. Unlike the situation which has been described for the fatty streak, the extent of intimal involvement by fibrolipid plaques appears to predeict the frequency and severity of the clinical manifestations of atherosclerosis in given population groups.[22-24]

In the aorta these plaques are seen most commonly in the abdominal portions of the vessel and often involve the mouths of intercostal and lumbar arteries. In such vessels as the common carotid artery it is rare to find lesions more advanced than fatty streaks, though calcified and ulcerated fibrolipid plaques are commonly seen in the region just beyond the point at which the common carotid branches and this modulation of the natural history of atherosclerotic lesions is, presumably, mediated by haemodyamic factors.

All fibrolipid plaques share two basic morphological components: a connective tissue cap which lies immediately beneath the endothelium and an underlying 'atheromatous pool' of lipid-rich, largely necrotic debris. Within these limits many morphological variants exist, such variants being very largely the expression of differences between the relative proportions of the connective tissue cap and the basal atheromatous pool and the complications related to such differences.

Fibrolipid plaques are elevated considerably above the surface of the surrounding non-involved intima. The media underlying the plaques often shows a significant degree of secondary thinning, but even when this is not present, the thickness of the plaque intima may exceed that of the media and adventitia combined. The morphological appearances of an individual lesion reflect to a very large extent the relative proportions of 'cap' and 'pool' within that lesion. In some lesions the proliferated connective tissue cap is the predominant element and this gives the plaque, as viewed from the surface, an opaque pearly white appearance. On viewing a section a yellow, lipid-rich base may be inconspicuous or even absent. Such firm plaques may cause luminal narrowing but are seldom associated with rupture and acute intra-lesional and intra-luminal thrombosis. In a study of advanced fibrous plaques within the femoral artery, the fibrous caps were found to consist of dense connective tissue in which spaces were noted that contained flat, pancake shaped cells showing the characteristics of smooth muscle. Electron microscopic examination showed that these muscle-cell containing lacunae consist of layers of basement membrane, collagen fibres and proteoglycans arranged in a concentric pattern.[25] In other lesions the basal accumulation of lipid, tissue debris and other blood-derived constituents may be massive and such a 'pool' is separated from the artery lumen only by a thin, easily ruptured layer of fibromuscular tissue. Such plaques are particularly at risk for major thrombotic events and this points to the importance of basal plaque necrosis as a key factor in the genesis of occlusive atherosclerosis-related disease. The surface ultrastructural changes occurring over plaques have been described recently.[19] Some lesions show variations in size, shape and orientation of surface cells such as to give rise to doubts as to whether they are truly endothelial. Ruffled, presumably activated macrophages can be seen beneath the endothelium and between adjacent endothelial cells. Denuded areas of intima ranging in area from less than a single endothelial cell to many cells are seen regularly and these are associated with the presence of adherent, activated platelets and, in some instances, with the presence of macrophages as well (Fig. 3).

Medial changes in relation to fibrolipid plaques

Medial thinning beneath fibrolipid plaques is quite common, and the extent and severity of this may play a part in determining whether an individual stenosis is 'fixed' or variable. Such changes in the media may reflect disturbances in its nutrition, this being derived normally from two sources. The adventitia and outer two-thirds of the media are supplied by diffusion from the vasa vasorum which course within the adventitia over the outer surface of the vessel. The intima and the inner one-third of the media are nourished by diffusion form the vessel lumen. The junction between these two zones constitutes a watershed situation and is the area of the artery wall most likely to suffer from disturbances of nutrition. A significant degree of intimal thickening, for example, will lead to

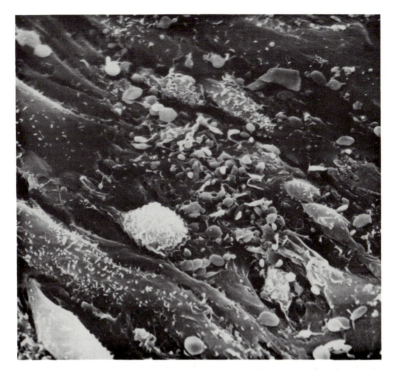

Fig. 3 Scanning electron micrograph of human coronary artery plaque. The intimal surface shows focal loss of endothelium, this being associated with the presence of numerous adherent platelets, many of which have undergone shape change. One or two of the macrophages are also adherent to the subendothelial structures (× 1,900).

a failure of lumen-derived nutrients to reach that depth within the vessel wall previously attainable. These ideas have received some support from histochemical studies showing that with increasing age and increasing intimal thickness, this mid-zone of the media is the site of a reduction in respiratory enzyme activity.[26]

Adventitial changes in relation to fibrolipid plaques

Fibrolipid plaques may be associated with three changes in the surrounding adventitial tissue.[27] These are:

an increase in fibrous tissue;
an increase in vascularity; and
the frequent presence of cellular aggregates consisting for the most part of B lymphocytes.

Adventitial lymphocytic aggregates were first described in the English literature as long ago as 1915.[28] The most extensive and systematic study of this phenomenon has been carried out by Schwartz and Mitchell.[29] Their findings indicated that plaque severity was the major determinant for this feature and that it was generally independent of age, sex and site within the arterial tree. The intensity of adventitial cellularity was found also to be related to the presence within the affected segments of recent thrombosis. Recent attempts to explain these appearances have focussed on the likelihood that such lymphocytic infiltrates are the expression of an immune response to the presence of oxidized lipids and ceroid in intra-lesional macrophages.[30]

Mural thrombosis in relation to fibrolipid plaques

Mural thrombi commonly occur over established plaques and these may become incorporated into the substance of the artery wall. In such circumstances, the degree of local intimal thickening may be increased, partly by the bulk of any unorganized residuum of the thrombus and partly because of a platelet-driven proliferative response. Much of the interest in a putative role for mural thrombosis as a contributor to plaque growth stems from the studies of the late J. B. Duguid[31–33] who put forward the view that many of the lesions classified as atherosclerotic are in fact altered thrombi which, by the process of organization, have been transformed into fibrous intimal thickenings. This, as Duguid pointed out, represents a partial return to the 'encrustation' hypothesis enunciated by Rokitansky a century before.[34]

In many instances it is easy to identify histologically transitions between obvious mural thrombi and sub-endothelial deposits or the remains of previously incorporated thrombotic material.[32,35,36] The use of immunohistochemical methods shows fibrin and fibrinogen to be present in many lesions[37,38] and the different morphological patterns of binding of antibodies raised against human fibrin may indicate whether fibrin or fibrinogen is present as a result of insudation or is a marker of incorporated mural thrombus.[38]

Carstairs[39] first identified platelet antigens within plaques using immunofluorescence. Subsequent studies[40,41] have shown that platelet antigens are found only in raised lesions and not in fatty streaks, and that their presence correlates with a pattern of antifibrin binding that suggests incorporation of thrombus. Such residua of thrombi may be found in up to 90% of the plaques in a single aorta and in up to 33% of coronary lesions.[42] There seems little reason to doubt, therefore, that frequent incorporation of mural thrombi occurs in relation to established atherosclerotic plaques.

The relation between these aortic plaques in which the immunohistological appearances suggest the presence of incorporated thrombus, to the presence or absence of ischaemic heart disease[43] shows a significant positive correlation. Whether this reflects an increased tendency to arterial thrombosis is unknown but is consistent with such a view. It is congruent with epidemiological evidence that raised plasma fibrinogen and Factor VII concentrations constitute independent risk factors for CHD.[44,45]

ACCUMULATION OF LIPID WITHIN THE ARTERIAL INTIMA

Like most mammalian tissues the artery wall contains a considerable amount of lipid which increases with normal growth and ageing.[46,47] The increase in the total lipid content of the artery wall with ageing is, in part brought about by a very slow and steady increase in free cholesterol, triglyceride and phospholipid. The concentration of esterified cholesterol within the vessel wall remains at a low level well into the second decade of life. It then starts to rise very rapidly, so that in normal arterial intima from patients aged 40–59, cholesterol ester makes up more than 40% of the total lipid content.

Atherosclerotic lesions contain far more lipid than does the normal intima. For example, fatty streaks contain about nine times as much lipid as does the lesion-free aortic intima in children and adolescents, and nearly four times as much as such intima in subjects aged between 40–59 years. Between 65–80% of this lipid is cholesterol, the ratio of esterified to free cholesterol being high in most lesions.[48] The cholesterol fatty acid pattern in juvenile fatty streaks differs from that seen in both plasma and fibro-lipid plaques, where the principal fatty acid is linoleic acid. Cholesterol esters in juvenile fatty streaks contain large amounts of oleate and relatively little linoleate.[49–51] The greater the number of 'foam cells' in the fatty streaks, the greater does this difference appear to be. This suggests that the cholesteryl ester of the fatty streak is derived from esterification of the cholesterol within the foam cells.

The cholesteryl esters accumulating within the arterial intima are believed to originate from the circulating lipoproteins which enter the interstitial space of the artery wall. This entry can be effected either by the formation of an ultrafiltrate of plasma or by transcytosis of LDL through the endothelial cells.

An origin for plaque cholesterol from plasma lipoprotein gains support from the morphological demonstration, by immunohistochemical methods, of apoproteins within atherosclerotic lesions.[53-57] The use of an elegant micro-immunoassay method[58] shows a high degree of correlation to exist between the amount of lipoprotein extractable from the artery wall and the plasma cholesterol concentration of the same patient.[47]

We now understand the fate of the lipoprotein entering the arterial intima to a much greater extent than previously. Much of this is due, in the first instance, to the discovery of the LDL receptor pathway by Brown and Goldstein,[59-62] and, in the second, to the recognition that the majority of foam cells in fatty streaks are macrophages.

Over 60% of the clearance of LDL from the plasma is mediated through its binding to the LDL receptor and subsequent endocytosis and most of this takes place as a result of ligand-receptor binding within the liver.

In the artery wall, however, available evidence suggests that the LDL receptor is not a major factor in the handling of the LDL which reaches the intima.[63] Patients suffering from the homozygous form of familial hypercholesterolaemia develop precocious and severe atherosclerosis and the Watanabe heritable hyperlipidaemic rabbit (WHHL), which is the homologue of FH, develops fatty streaks rich in foam cells even though it is totally devoid of LDL receptors.[8,64] In addition normal monocytes and monocyte-derived macrophages in culture are not converted into foam cells by incubation with unmodified low density lipoprotein.[65,66] However, chemically modified LDL is avidly taken up by monocytes in cell culture. This is believed to be due to the existence of other receptors which have been called 'scavenger receptors'. These do not recognize unmodified LDL but bind LDL which has been either acetylated or conjugated with malondialdhyde, an aldehyde product of lipid peroxidation.[67]

Production of oxidatively modified LDL

If LDL is incubated in the presence of cultured endothelial cells, arterial smooth muscle cells or macrophages it undergoes extensive changes.[68-70] The most striking of these is the ability to bind to the 'scavenger receptor' on macrophages and thus to undergo endocytosis and intracellular degradation. In a mirror image fashion, the modified LDL loses its ability to be recognized by the LDL receptor. In addition the modified LDL becomes strongly chemoattractant for human monocytes, this being due to alterations in the lipid moiety of the LDL, in particular the accumulation of significant amounts of lysylphosphatidylcholine.[71-73] Not only can the altered LDL attract monocytes but it also inhibits the basal motility of mouse peritoneal macrophages and their ability to respond to other chemoattractants. Since many studies of the effects of hyperlipidaemia on the artery wall suggest that one of the earlier events, in morphological terms, is the adhesion of white cells to the endothelial surface, followed by entry into the subjacent intima, this chemotactic effect of modified LDL may be a primary event in the genesis of the fatty streak[63,74] (Fig. 4).

Nature of the modification of LDL

The changes that occur in the LDL molecule, whether following incubation with one of the cell types mentioned above or incubation with plasma containing copper or iron, depend on peroxidation of polyunsaturated fatty acids in the lipid fraction and thus on free radical generation. This is shown by the fact that LDL modification by endothelial cells is blocked if the incubation takes place in the presence of **antioxidants** such as vitamin E or butylated hydroxytoluene. Modification of LDL involves changes in both the lipid and the protein moieties. In the lipid there is a significant degree of conversion of lecithin to lysolecithin and the cytoxic effects of modified LDL on endothelial cells in culture probably depends largely on this.[75,76] However in the course of peroxidation of unsaturated lipid moieties, 4-hydroxyalkenals can be generated from omega-6- and omega-3 polyunsaturated fatty acids and these biologically active compounds can cause severe disturbances of cell function at a number of levels.[77] Steinbrecher[78] has shown that cleavage products of arachidonic acid can react with apoprotein B and this reaction involves blocking of the epsilon-amino groups of lysine residues. It is now believed that the interaction of aldehydic lipid peroxidation products with these epsilon-amino groups of lysine residues produces new epitopes which are recognized by the scavenger receptor on macrophages.[79] Thus the generation of free radicals in the subendothelial space may have a variety of important results.

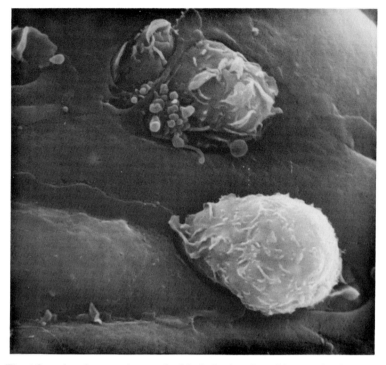

Fig. 4 Scanning electron micrograph of the intimal surface of the aorta in a hyperlipidaemic rabbit showing two macrophages. One is adherent to the plasma membrane of the endothelium; the other appears to have in part gained entry to the subendothelial space and is traversed by a 'bridge' of endothelial cytoplasm (× 4,900).

Cytotoxic chemical species can be formed which may play a role in the later stages of plaque evolution; foam cell formation may be mediated through the operation of the scavenger receptor and monocytes may be attracted to the relevant area of intima and immobilized there.

Oxidation of low density lipoprotein 'in vivo'

It is important to know whether oxidative modification of LDL occurs **in vivo**. If changes in LDL play a role in atherogenesis, then the use of appropriate antioxidants at appropriate dose levels should inhibit the genesis of atherosclerosis in hyperlipidaemic animals. This has been demonstrated by two groups who showed that administration of the drug Probucol, which has a mildly hypolipidaemic action but which is also an antioxidant, to Watanabe rabbits resulted in a marked degree of inhibition of lesion development.[80,81] This appeared to be largely independent of any effect on plasma lipid concentrations.

CONNECTIVE TISSUE PROLIFERATION AND THE ARTERIAL SMOOTH MUSCLE CELL

Lumenal stenosis is due to a striking focal increase in the amount of intimal connective tissue. The cell type responsible for this process is the intimal smooth muscle cell (SMC). Amongst the many interesting characteristics of arterial smooth muscle perhaps the most fascinating is its ability to exist in more than one phenotype. At one extreme is a smooth muscle cell the function of which, like most other muscle cells, is almost entirely that of contraction; at the other extreme is a cell almost exclusively concerned with proliferation and with synthesis of a number of extracellular tissue components.

The expression of altered phenotype is controlled by the interaction of a series of chemical signals. These may exert their effect either by directly stimulating the smooth muscle cells by a receptor ligand interaction, or by the removal of some inhibitory influence(s) such as heparin. In their alternate, non-contractile phenotype, SMC cells can proliferate to a striking degree in a response to local increases in growth factors. These may be released in a paracrine fashion from nearby endothelial cells, platelets or macrophages or, in an autocrine fashion, from smooth muscle cells themselves. The behaviour of such cells has been studied in a number of models of arterial injury, following which medial smooth muscle cells migrate to the intima and proliferate there. Other features of the altered smooth muscle cell include the expression of plasma membrane receptors for low density lipoprotein, and the ability to secrete a number of prostanoids. The source of the new extracellular connective tissue formed both in the course of atherogenesis and in the response to arterial injury is almost certainly the smooth muscle cell in its synthetic phenotype.[82] Compelling evidence exists that medial smooth muscle cells derived by explant culture from non-human primate aorta produce not only soluble elastin but large amounts of a glycoprotein with an amino-acid composition identical with that found in the micro-fibrillar component of intact elastic fibres.[83,84] Ross[85] has found that cultured arterial smooth muscle cells can also secrete collagen and this has been confirmed by McCullagh and Balian.[86] Arterial smooth muscle cells in culture also show the ability to produce glycosoaminoglycans, dermatan sulphate being, quantitatively, the most prominent. Since a major portion of the connective tissue cap of a fibro-lipid plaque consists of just these components, the significance of the **ex vivo** findings appears considerable and it seems reasonable to assume that modulation of the arterial smooth cells from the '**contractile**' to the '**synthetic**' phenotype is responsible for the connective tissue proliferation which is a key element in the evolution of atherosclerosis.

Role of growth factors in controlling the intimal smooth muscle population

We owe our understanding of the effect of growth factors on vascular smooth muscle in the first instance to Russell Ross and his colleagues. They showed that the mitogenic effect of serum derived from whole, clotted blood on arterial smooth muscle cells in culture is largely due to a factor released from platelets and thus called **platelet-derived growth factor (PDGF)**.[87-92] The term **platelet-derived growth factor** is something of a misnomer. It is certainly stored in and released from the alpha granules of platelets (together with platelet factor IV and beta-thromboglobulin) but can also be produced by endothelial cells, macrophages and arterial smooth muscle cells in their synthetic phenotype. The dimeric protein is composed of two homologous polypeptide chains A and B and the genes encoding these have been mapped to different chromosomes, their expression thus being independently regulated.[93,94] It binds with high affinity to receptors on arterial smooth muscle cells, fibroblasts 3T3 cells (an immortalized line of murine fibroblasts), and other mesenchymal cells though there is no evidence of an ability to bind to arterial endothelium. It differs in an important respect from all other growth factors thus far discovered in that it is not only mitogenic but is also chemoattractant and this may account for the migration of smooth muscle cells from the arterial media which is seen in atherogenesis and in the response to arterial injury.

PDGF is a **competence** type of growth factor bringing cells to which it binds out of G_0 and into the cell cycle rather than a **progression** factor which has a mitogenic effect only on cells already in cycle. In so doing, it causes activation of the phosphatidylinositol pathway, the transcription of c-myc and c-fos proto-oncogenes and an increase in both cytoplasmic calcium concentration and pH. Its B chain shows an 87% degree of homology with the gene product of the transforming oncogene of the simian sarcoma virus (**v-sis**) and, more importantly, with that of the cellular proto-oncogene **c-sis**.[95,96] Expression of this gene is almost certainly an important element in wound repair and the cells of several malignant tumours have been shown to secrete PDGF which, presumably, stimulates proliferation of these cells in an autocrine fashion since treatment of such cells in culture with antibodies raised against PDGF blocks the incorporation of tritiated thymidine into the tumour cells.[97]

Expression of growth factors in human atherosclerotic lesions

Barrett and Benditt[98] have found elevated levels of mRNA for PDGF in atherosclerotic plaques as compared with normal artery wall. The same workers found that PDGF-A chain seems to be expressed by the smooth muscle cells of the normal media at about the same level as in lesions. mRNA for both the A and B chains of PDGF has been identified in plaque endothelium and also in cells within the intima described as 'mesenchymal' cells and which appear to be transcribing mRNA for the A chain in particular but which did not bind any of the available cell-specific markers.[99] In this study PDGF mRNA was not found in plaque macrophages though Barrett and Benditt[100] found that in atherosclerotic plaques PDGF B chain mRNA levels correlated strongly with **fms** mRNA levels (**fms** is the cellular proto-oncogene which codes for the colony stimulating factor-1 receptor and is regarded by some workers as a cell specific marker for macrophages).

Hybridization with a probe for the PDGF receptor showed many of the cells in the plaques to be expressing this entity. Almost all of these cells were within the intima and showed the same mesenchymal morphology as described above. No cells resembling endothelium or lymphocytes and almost no foam cells or haemosiderin-containing cells were positive for the PDGF receptor probe. Those cells which did show mRNA for the PDGF receptor did not react with antibody markers for endothelial cells, lymphocytes or monocytes/macrophages. Few receptor positive cells were found in the media of the arterial samples examined, an interesting observation, since, as Wilcox and his colleagues[99] point out, the differential expression of the PDGF receptor within the intima and the media, 'establishes the presence of a target in the part of the vessel wall known to undergo selective proliferation in atherosclerosis and following intimal injury such as occurs in the course of angioplasty.'

These data suggest that the chemical regulation of smooth muscle cell migration, proliferation and function in the arterial wall and the connective tissue increase which results from ligand-receptor induced changes in smooth muscle phenotype are likely to be very complex phenomena *in vivo*. The situation at this time has been admirably summarized by Barrett and Benditt[100]: 'the findings are consistent with the possibility that the regulation of smooth muscle cell proliferation in atherosclerosis is multifactorial, involving inducers and suppressors, and could centre on a "cytokine network" of intercellular signalling factors, including immune modulators and peptide growth factors, as well as other substances.'

Clonality of smooth muscle cells in atherosclerotic lesions

The data presented above suggest that these smooth muscle cell responses are, in a broad sense, a **reaction to injury**.[91] An alternate view has been proposed by Benditt and Benditt.[101] They showed that in many instances the smooth muscle population within atherosclerotic plaques is monotypic and, they believe, monoclonal. They suggest, therefore, that the smooth muscle cell proliferation occurring in atherogenesis is more closely allied to the cell proliferation seen in neoplasia than in that of the repair phase that follows injury.

It is of course possible that the monotypic nature of the cell population in a plaque might be explained on a basis other than that of monoclonality. Some alternative mechanisms have been canvassed including the possibility that the enzyme variant itself or some gene linked by location on the X chromosome confers some selective advantage on the cells; the possibility that lesion-free clusters of some cells are so large that, if multiplication were triggered, the resulting mass would be monotypic and the possibility that repetitive sampling of the cell population during cycles of cell multiplication and cell death might lead to a drift towards a monotypic population.[102] This last suggestion gains some support from the studies of Zavala and his associates[103] who cultured fibroblasts from 29 Negro females heterozygous for G-6-PD through multiple passages. In 19 of the 29 the culture revealed only one phenotype. In the remaining 10 the cultures died out while still ditypic.

If the monotypic nature of the cell population in some plaques is indeed a reflection of monoclonal smooth muscle cell proliferation then it seems not unlikely that the trigger for this proliferation is a mutagen. In human plasma, potentially mutagenic hydrocarbons are carried in the lipoprotein fraction[104] and at least two samples of human aorta have been shown to contain a mixed function oxidase system (aryl-4-mono-oxygenase) which might transform premutagens into substances that are cytotoxic or which might attach themselves covalently to DNA and modify it.[105]

Experimental evidence to support this view of the genesis of some plaques is not lacking. Herpes virus mRNA has been shown to be present by hybridization in cells of human atherosclerotic plaques[106]; weekly injections of carcinogenic hydrocarbons have resulted in the appearance of fibro-muscular plaques in the abdominal aortae of cockerels[107] and 15 week old cockerels, which had been injected at the age of 4 days with the oncogenic Marek's disease virus, developed focal plaques in the thoracic aorta.[108,109]

An experiment of great potential significance in relation to the monoclonal hypothesis has been the demonstration by Penn and his colleagues[107] that DNA extracted from samples of some human coronary artery plaques contained a transforming gene. The assay used relies on the incorporation and expression of dominant transforming DNA sequences by NIH 3T3 cells, an immortalized murine fibroblast line which has proved susceptible to transformation by genes of the **ras** group. The morphological correlate of such transformation is the appearance in monolayer cultures of the 3T3 cells of foci where the cells are heaped up (presumably having lost contact inhibition) and where the individual cells are smaller and rounder than the surrounding monolayer. DNA from such foci must be able to produce the same changes in other 3T3 cultures and the final link in proving that transformation has actually occurred, is provided by assessing the ability of cells from the 'foci' to produce tumours when injected into suitable hosts, most notably the athymic 'nude' mouse. Three groups of human coronary artery plaque DNA samples gave rise to transformed foci when transfected into 3T3 cells, while DNA samples from a variety of non-malignant tissues failed to do so. Southern blotted DNA from the transformed foci yielded positive signals when hybridized to a ^{32}P-labelled nick-translated repetitive, human **Alu** sequence indicating the presence of human genetic material, but failed to show any active **ras** oncogenes. Primary focus cells from each of five clones produced slowly growing tumours in 6/42 nude mice and samples of the tumour DNA also hybridized to **Alu**. These data suggest that somatic gene alterations can play a significant role in the genesis of some atherosclerotic plaques.

A POSSIBLE ROLE FOR THE MACROPHAGE IN RELATION TO PLAQUE NECROSIS

Necrosis at the plaque base is a key event in the natural history of atherosclerosis being associated with an increased risk of plaque rupture and consequently of acute thrombosis. The possibility that such basal necrosis might be due, at least in part, to the release of active chemical species by plaque macrophages is of considerable interest. Activated macrophages can, for instance, release oxygen free-radicals and, in the ex vivo situation, such release is potentiated by exposure of macrophages to oxidized LDL.[110] Macrophages can also be the source of powerful proteases, including collagenase and elastase, either by reverse endocytosis or following the death of the macrophages themselves. These observations are derived largely from cell culture studies, and, therefore, must be regarded with some caution, but there is sufficient evidence to make a **prima facie** case for more extensive investigation of the role of the macrophage in the later stages of the natural history of atherosclerosis.

REFERENCES

1 Crawford T. The Pathology of Ischaemic Heart Disease. London: Butterworth's, 1977; p. 32
2 Davies MJ, Thomas AC. Pathological basis and microanatomy of occlusive thrombus formation in human coronary arteries. In: Born GVR, Vane JR, eds. Interactions between platelets and vessel walls. R Soc London, 1981: pp. 9–12
3 Falk E. Plaque rupture with severe pre-existing stenosis precipitating coronary thrombosis: characteristics of coronary atherosclerotic plaques underlying fatal occlusive thrombi. Br Heart J 1983; 50: 127–134
4 Davies MJ, Thomas AC. Thrombosis of acute coronary artery lesions in sudden ischaemic death. N Engl J Med 1984; 310: 1137–1140
5 Haust MD. The morphogenesis and fate of potential and early atherosclerotic lesions in man. Hum Pathol 1971; 2: 1–30
6 Gerrity RG, Naito HK, Richardson M, Schwartz CJ. Dietary induced ather-

ogenesis in swine: morphology of the intima in prelesion stages. Am J Pathol 1979; 95: 775

7 Faggiotto A, Ross R. Studies of hypercholesterolemia in the non-human primate: II. Fatty streak conversion to fibrous plaque. Arteriosclerosis 1984; 4: 341

8 Rosenfeld ME, Tsukada T, Fown' AM, Ross R. Fatty streak initiation in Watanabe heritable hyperlipemic and comparable hypercholesterolemic fat-fed rabbits. Arteriosclerosis 1987; 7: 9–23, 1987

9 Seddon AM, Woolf N, LaVille A et al. Hereditary hyperlipidaemia and atherosclerosis in the rabbit due to overproduction of lipoproteins. II. Preliminary report of arterial pathology. Arteriosclerosis 1987; 7: 113–124

10 Taylor RG, Lewis JC. Endothelial cell proliferation and monocyte adhesion to atherosclerotic lesions of white Carneau pigeons. Am J Pathol 1986; 125: 152–160

11 Schwartz CJ, Ardlie NG, Carter RF, Paterson JC. Gross aortic sudanophilia and hemosiderin deposition. A study on infants, children and young adults. Arch Pathol 1967; 83: 325–332

12 Caro CG, Fitzgerald JM, Schroter RC. Arterial wall shear and distribution of early atheroma in man. Nature 1969; 223: 1159–1160

13 Caro CG, Fitzgerald JM Schroter RC. Atheroma and arterial wall shear. Observation, correlation and proposal of a shear dependent mass transfer mechanism for atherogenesis. Proc R Soc Lond 1971; 177: 109–159

14 Chien S. Significance of macrorheology and microrheology in atherogenesis. Ann NY Acad Sci USA 1976; 275: 10–27

15 Fowler S. Characterization of foam cells in experimental atherosclerosis. In: Proceedings of the 5th Paavo Nurmi Symposium: Thrombosis and Blood Vessel Wall Interaction in Coronary Heart Disease. Acta Med Scand (Suppl) 1980; 642: 151–158

16 Mitchinson MJ. Macrophages, oxidised lipids and atherosclerosis. Med Hypoth 1983; 12: 171–178

17 Klurfeld DM. Identification of foam cells in human atherosclerosis lesions as macrophages using monoclonal antibodies. Arch Pathol Lab Med 1985; 109: 445–449

18 Mitchinson MJ, Ball RY. Macrophages and atherogenesis. Lancet 1987; ii: 146–149

19 Davies MJ, Woolf N, Rowles PM, Pepper J. Morphology of the endothelium over athosclerotic plaques in human coronary arteries. Br Heart J 1988; 60: 459–464

20 McGill HC Jnr. The lesion. In: Atherosclerosis III. Proc Third Int Symposium. Schettler G, Weizel A; Berlin: Springer-verlag, 1974: pp. 27–38

21 Restrepo C, Tracy RE. Variation in human aortic fatty streaks among geographic locations. Atherosclerosis 1975; 21: 179–193

22 Tejada C, Strong JP, Montenegro MR et al. Distribution of aortic and coronary atherosclerosis by geographic location, race and sex. Lab Invest 1968; 18: 509–526

23 Strong JP, Solberg LA, Restrepo C. Atherosclerosis in persons with coronary heart disease. Lab Invest 1968; 18: 527–537

24 Deupree RH, Fields RI, McMahan CA, Strong JP. Atherosclerotic lesions and coronary heart disease. Key relationships in necropsied cases. Lab Invest 1973; 28: 252–262

25 Ross R, Wight TN, Strandness E, Thiel B. Human atherosclerosis I. Cell constitution and characterization of advanced lesions of the superficial femoral artery. Am J Pathol 1984; 114: 79–93

26 Adams CWM, Bayliss OB. The relationship between diffuse intimal thickness, medial enzyme failure and intimal lipid deposit in various human arteries. J Ath Rev 10: 327–339, 1969

27 Mitchell JRA, Schwartz CJ. The planometric assessment of aortic disease severity. In: Arterial Disease; Oxford: Blackwell 1965; 386–396
28 Albutt C. Diseases of the arteries, including angina pectoris. London: MacMillan 1915; 1: 468
29 Schwartz CJ, Mitchell JRA. Cellular infiltration of human arterial adventitia associated with atheromatous plaques. Circulation; 26: 73
30 Parums D, Mitchinson MN. Demonstration of immunoglobulin in the neighbourhood of advanced atherosclerotic plaques. Atherosclerosis 1981; 38: 211–216
31 Duguid JB. Thrombosis as a factor in the pathogenesis of coronary atherosclerosis. J Pathol Bacteriol 1946; 58: 207–212
32 Duguid JB. Thrombosis as a factor in the pathogenesis of aortic atherosclerosis. J Pathol Bacteriol 1948; 60: 57–61
33 Duguid JB. The Dynamics of Atherosclerosis . Aberdeen: Aberdeen University Press 1976: pp. 44–49
34 Rokitansky C. In: Handbuch der Pathologisches Anatomie, Braunmuller and Seidel, 1844: vol 2
35 Heard BE. An experimental study of thickening of the pulmonary arteries of rabbits produced by organisation of fibrin. J Pathol Bacteriol 1949; 64: 13–19
36 Morgan AD. The Pathogenesis of Coronary Occlusion. Oxford: Blackwell, 1956
37 Woolf N, Crawford T. Fatty streaks in aortic intima studied by an immunohistochemical technique. J Pathol Bacteriol 1960; 80: 405–408
38 Woolf N. The distribution of fibrin within the aortic intima: an immunohistochemical study. Am J Pathol 1961; 39: 521–532
39 Carstairs KC. The identification of platelets and platelet antigens in tissue sections. J Pathol Bacteriol 1965; 90: 225–231
40 Woolf N, Carstairs KC. The survival time of platelets in experimental mural thrombi. J Pathol 1967; 97: 595–601
41 Hudson J, McCaughey WTE. Mural thrombosis and atherogenesis in coronary arteries and aorta. Atherosclerosis 1974; 19: 543–553
42 Chandler AB, Terrell Pope J. Arterial thrombosis in atherogenesis. In: Hautvast JGAJ, Hermus RJJ, van der Haap F (eds). Blood and arterial wall in atherogenesis and arterial thrombosis. Iifma Scientific Symposium No 4; Brill, Leiden, 1974; 111–118
43 Woolf N, Sacks MI, Davies MJ. Aortic plaque morphology in relation to coronary artery disease. Am J Pathol 1969; 57: 187–197
44 Meade TW, North WRS, Chakrabarti R et al. Haemostatic function and cardiovascular death: early results of a prospective study. Lancet 1980; i: 1050–1054
45 Meade TW. Thrombogenic factors. In: Olsson AG, ed. Atherosclerosis: Biology and Clinical Science. Edinburgh: Churchill Livingstone 1987; pp. 453–455
46 Smith EB. The influence of age and atherosclerosis on the chemistry of the aortic intima. I. The lipids. J Ath Res 1965; 5: 224–240
47 Smith EB. The relationship between plasma and tissue lipids in human atherosclerosis. Adv Lipid Res 1974; 12: 1
48 Bottcher CJF. Phospholipids of atherosclerotic lesions in human aorta. In: Jones RJ, (ed.) Evolution of the atherosclerotic plaque. Chicago: University Press 1964; 109–116
49 Bottcher CJF, Woodford FP, Romeny-Wachter CTH et al. Fatty acid distribution in lipids of the aortic wall. Lancet 1960; i: 1378–1383
50 Geer JC, Malcolm GT. Cholesteryl ester fatty acid composition of human aorta fatty streaks and normal intima. Exp Mol Pathol 1964; 4: 500–507
51 Smith EB, Evans PH, Downham MD. Lipid in the aortic intima. The corre-

lation of morphological and chemical characteristics. J Ath Res 1967; 7: 171–186

52 Newman HAI, Zilversmit DB. Quantitative aspects of cholesterol flux in rabbit atheromatous lesions. J Biol Chem 1962; 237: 2078–2084

53 Watts HF. Pathogenesis of human coronary artery atherosclerosis. Demonstration of serum lipoproteins in the lesions and of localized intimal enzyme defects by histochemistry. Circulation 1961; 24: 1066

54 Watts HF. Role of lipoproteins in the formation of atherosclerotic lesions. In: Jones RJ ed. Evaluation of the Atherosclerotic Plaque. Chicago: University Press 1963; 117–132

55 Kao VCT, Wissler RW. A study of the immunochemical localization of serum lipoproteins and other plasma proteins in human atherosclerotic lesions. Exp Mol Pathol 1965; 4: 465–479

56 Woolf N, Pilkington TRE. The immunohistochemical demonstration of lipoproteins in vessel walls. J Pathol Bacteriol 1965; 91: 459–463

57 Walton KW, Williamson N. Histological and immunofluorescent studies in the evolution of the human atheromatous plaque. J Ath Res 1968; 8: 599–624

58 Smith EB, Slater RS. Relationship between low density lipoprotein in aortic intima and serum-lipid levels. Lancet 1972; i: 463–469

59 Brown MS, Goldstein JL. Receptor-mediated control of cholesterol metabolism. Science 1976; 191: 150–154

60 Goldstein JL, Brown MS. The low density lipoprotein pathway and its relationship to atherosclerosis. Annu Rev Biochem 1977; 46: 897–930

61 Brown MS, Kovanen PT, Goldstein JL. Regulation of plasma cholesterol metabolism. Science 1981; 191: 150–154

62 Brown MS, Goldstein JL. How LDL receptors influence cholesterol and atherosclerosis. Sci Am 1984; 251: 58–66

63 Steinberg D, Parthasarathy S, Carew TE et al. Beyond cholesterol. Modifications of low-density lipoprotein that increase its atherogenicity. N Engl J Med 1989; 320: 915–924

64 Buja LM, Kita T, Goldstein JL et al. Cellular pathology of progressive atherosclerosis in the WHHL rabbit. An animal model of familial hypercholesterolemia. Arteriosclerosis 1983; 3: 87–101

65 Goldstein JL, Ho YK, Basu SK, Brown MS. Binding site on macrophages that mediate uptake and degradation of acetylated low density lipoprotein, producing massive cholesterol deposition. Proc Natl Acad Sci USA 76: 333–337, 1979

66 Brown MS, Goldstein JL. Lipoprotein metabolism in the macrophage: implications for cholesterol deposition in atherosclerosis. Annu Rev Biochem 52: 223, 1983

67 Fogelman AM, Schechter I, Seager J et al. Malondialdehyde alteration of low density lipoproteins leads to cholesteryl ester accumulation in human monocyte-macrophages. Proc Natl Acad Sci USA 1980; 77: 2214

68 Henrickson T, Mahoney EM, Steinberg D. Enhanced macrophage degradation of low density lipoprotein previously incubated with cultured endothelial cells: recognition by receptors for isolated low density lipoproteins. Proc Natl Acad Sci USA 1982; 78: 6499–6503

69 Henrikson T, Mahoney EM, Steinberg D. Enhanced macrophage degradation of biologically modified low density lipoprotein. Arteriosclerosis 1983; 3: 149–159

70 Parthasarathy S, Young SG, Witzum JL et al. Probucol enhances oxidative modification of low density lipoprotein. J Clin Invest 1986; 77: 641–644

71 Quinn NT, Parthasarathy S, Steinberg D. Endothelial cell-derived chemotactic activity for mouse peritoneal macrophages and the effect of modified forms of low-density lipoprotein. Proc Natl Acad Sci USA 1985; 82: 5949

72 Quinn MT, Parthasarathy S, Steinberg D. Lysophosphatidylcholine: a chem-

otactic factor for human monocytes and its potential role in atherogenesis. Proc Natl Acad Sci USA 1988; 85: 2805–2809

73 Quinn MT, Parthasarathy S, Fong LG, Steinberg D. Oxidatively modified low density lipoproteins: potential role in recruitment and retention of monocyte/macrophage during atherogenesis. Proc Natl Acad Sci USA 1987; 84: 2995–2998

74 Steinberg D. Metabolism of lipoproteins and their role in the pathogenesis of atherosclerosis. In: Stokes J III, Mancini M, eds. Atherosclerosis Reviews 1988; volume 18. New York: Raven Press

75 Cathcart MK, Morel DW, Chisholm GM III. Monocytes and neutrophils oxidise low density lipoprotein making it cytotoxic. J Leucocyte Biol 1985; 38: 341

76 Henrikson T, Evensen SA, Carlander B. Injury to human endothelial cells in culture induced by low density lipoproteins. Scand J Clin Lab Invest 1979; 39: 361

77 Estebauer H, Zollner H, Schaur RJ. Hydroxyalkenals: cytotoxic products of lipid peroxidation. ISI Atlas of Science: Biochemistry 1988; 1: 311–317

78 Steinbrecher UP. Oxidation of human low density results in derivatization of lysine residues of apolipoprotein B by lipid peroxide decomposition products. J Biol Chem 1987; 262: 3603–3608

79 Estebauer H, Quehenberger O, Jurgens G. Effect of peroxidative conditions on human plasma low-density lipoproteins. In: Nigam et al., eds. Lipid Peroxidation and Cancer. Berlin: Springer-Verlag, 1988: pp. 203–213

80 Carew TE, Schwenke DC, Steinberg D. Antiatherogenic effect of probucol unrelated to its hypercholesterolaemic effect: evidence that antioxidants in vivo can selectively inhibit low density lipoprotein degradation in macrophage-rich fatty streaks and slow the progression of atherosclerosis in the Watanabe heritable hyperlipidemic rabbit. Proc Natl Acad Sci USA 1987; 84: 7725–7729

81 Kita T, Nagaro Y, Yukode M et al. Probucol prevents the progression of atherosclerosis in Watanabe heritable hyperlipidemic rabbit, an animal model for familial hypercholesteroleamia. Proc Natl Sci USA 1987; 84: 5928

82 Burke JM, Ross R. Synthesis of connective tissue macromolecules by smooth muscle. Int Rev Connect Tissue Res 1979; 8: 119–157

83 Narayanan AS, Sandberg LE, Ross R, Layman DL. The smooth muscle cell. III. Elastin synthesis in arterial smooth muscle cell culture. J Cell Biol, 68: 411–417, 1976

84 Ross R. Platelets, smooth muscle proliferation and atherosclerosis. Acta Med Scand 1980; (Suppl 642): 49–54

85 Ross R. The smooth muscle cell in connective tissue metabolism and atherosclerosis. In: Kulonen E, ed. The Biology of the Fibroblast Proc Sigrid Juselius Foundation Symposium, Turku, Finland. London: Academic Press 1973; 623–636

86 McCullagh KE, Balian G. Collagen characterization and cell transformation in human atherosclerosis. Nature 1975; 258: 73–75

87 Ross R, Glomset J, Kariya B, J, Harker L. A platelet-dependent serum factor that stimulates the proliferation of arterial smooth muscle cells in vitro. Proc Natl Acad Sci USA 1974; 71: 1207

88 Rutherford RB, Ross R. Platelet factors stimulate fibroblasts and smooth muscle cells quiescent in plasma serum to proliferate. J Cell Biol 1976; 69: 196–200

89 Ross R, Glomset J, Harker L. Response to injury and atherogenesis. Am J Pathol 1977; 86: 675–684

90 Ross R, Harker L. Platelets, endothelium and smooth muscle cells in atherosclerosis. In: Day HJ, Molony BA, Nishizawa EE, Rynbarth RH eds. Thrombosis: Animal and Clinical Models. New York: Plenum Press, 1978; pp. 125–144

91 Ross R. The pathogenesis of atherosclerosis—an update. N Engl J Med 1986; 314: 488–501

92 Ross R. Cellular interactions in atherosclerosis—the role of growth factors. In: Crepaldi G, Gotto AM, Manzato E, Baggio G, eds. Atherosclerosis VIII. Amsterdam: Excerpta Medica 1989; pp. 13–19

93 Betzholtz C, Johnsson A, Aildin C-H et al. c-DNA and chromosomal localisation of human platelet derived growth factor A-gene and its expression in tumour cell lines. Nature 1986; 320: 695–699

94 Swan DC, McBride DW, Robbins KC et al. Chromosomal mapping of the simian sarcoma virus oncogene analogue in human cells. Proc Natl Acad Sci USA 1982; 79: 4691–4695

95 Waterfield MD, Scrace GT, Whittle H et al. Platelet-derived growth factor is structurally related to the putative transforming protein p28^{v-sis} of simian sarcoma virus. Nature 1983; 304: 35–39

96 Doolittle RF, Hunkapiller MW, Hood LE et al. Simian sarcoma virus oncogene, v-sis, is derived from the gene (or genes) encoding a platelet-derived growth factor. Science 1983; 221: 275–277

97 Sporn MB, Roberts AB. Autocrine growth factors in cancer. Nature 1985; 313: 745

98 Barrett TB, Benditt EP. sis (PDGF-beta chain) gene transcript levels are elevated in human atherosclerotic lesions compared to normal artery. Proc Natl Acad Sci USA 1987, 84: 1099

99 Wilcox JN, Smith KM, Williams LT et al. Platelet derived growth factor mRNA detection in human atherosclerotic plaques by in situ hybridization. J Clin Invest 1988; 82: 1134–1143, 1988

100 Barrett TB, Benditt EP. Platelet-derived growth factor gene expression in human atherosclerotic plaques and normal artery wall. Proc Natl Acad Sci USA 1988; 85: 2810–2814

101 Benditt EP, Benditt JM. Evidence for a monoclonal origin of human atherosclerotic plaques. Proc Natl Acad Sci USA 1973; 70: 1753–1756

102 Thomas WA, Florentin RA, Reiner JM et al. Alterations in population dynamics of arterial smooth muscle cells during atherogenesis. IV. Evidence for a polyclonal origin of hypercholesterolaemic diet-induced atherosclerotic lesions in young swine. Exp Mol Pathol 1976; 24: 244–260

103 Zavala C, Herner G, Fialkow PJ. Evidence for selection in cultured diploid fibroblast strains. Exp Cell Res 1978; 177: 137

104 Benditt EP. Implications of the monoclonal character of human atherosclerotic plaques. Beitrage zur Pathologie 1976; 158: 433–444

105 Benditt EP. The artery wall and the environment. In: Fifth Paavo Nurmi Symposium. Thrombosis and blood-vessel wall interactions in coronary heart disease. Acta Med Scand, 1980 (unpublished)

106 Benditt EP, Barrett T, McDougall JK. Viruses in the etiology of atherosclerosis. Proc Natl Acad Sci USA 1983; 80: 6386

107 Penn A, Garte SJ, Warren L et al. Transforming gene in human atherosclerotic plaque DNA. Proc Natl Acad Sci USA 1986; 83: 7951

108 Fabricant CG, Fabricant J, Litrenta MM, Minick CR. Virus-induced atherosclerosis. J Exp Med 1978; 148: 335–340

109 Fabricant CG. The consequence of infection with a herpes virus. Adv Vet Sci Comp Med 1985; 30: 39

110 Hartung H-P, Kladetsky RG, Melnik B, Hennerici M. Stimulation of the scavenger receptor on monocytes-macrophages evokes release of arachidonic acid metabolites and reduced oxygen species. Lab Invest 1986; 55: 209–216

British Medical Bulletin (1990) Vol. 46, No. 4, pp. 986–1004

Primary hyperlipidaemia

G R Thompson

MRC Lipoprotein Team, Hammersmith Hospital, London, UK

It is estimated that over 60% of the variability in serum lipids is genetically determined, most of this variation being due to polygenic influences. Interaction between the latter and environmental factors is probably the commonest cause of hyperlipidaemia in the general population.

Familial forms of hyperlipidaemia are usually more clearly defined, especially those which have a monogenic or dominant pattern of inheritance, but are less common. This type of disorder, exemplified by familial hypercholesterolaemia, is expressed independently of environmental influences. In contrast, in familial type III hyperlipoproteinaemia inheritance of the underlying gene defect is often insufficient to produce hyperlipidaemia unless additional environmental or genetic influences co-exist. Rarely, hyperlipidaemia is recessively inherited, as in familial deficiency of lipoprotein lipase and of apolipoprotein CII.

Primary hyperlipidaemias characterized by severe hypertriglyceridaemia predispose to acute pancreatitis whereas those disorders characterized by hypercholesterolaemia, apart from hyperαlipoproteinaemia, are associated with an increased risk of premature vascular disease.

The World Health Organization (WHO) classification of lipoprotein phenotypes[1] provides a useful means of indicating which lipoproteins are present in excess in individual patients but has the major limitation that it does not differentiate between primary and secondary forms of hyperlipidaemia. This distinction depends upon demonstrating the presence or absence of underlying causes and the results of family studies.

0007–1420/90/0046–0986/$10.00

The type I phenotype indicates hypertriglyceridaemia due to chylomicronaemia; in type IIa the hypercholesterolaemia reflects an increase in low density lipoprotein (LDL) cholesterol which in type IIb is accompanied by mild to moderate hypertriglyceridaemia, due to an increase in very low density lipoprotein (VLDL); in type III serum cholesterol and triglyceride are both raised, due to accumulation of chylomicron and VLDL remnants; in type IV fasting hypertriglyceridaemia and mild to moderate hypercholesterolaemia are due to an increase in VLDL but the LDL-cholesterol is normal; and in type V marked hypertriglyceridaemia is due both to chylomicronaemia and an increase in VLDL.

As already stated each phenotype can represent either primary or secondary hyperlipidaemia. Frequently, primary hyperlipidaemia is polygenically-determined and therefore rather ill-defined but several monogenic or dominantly-inherited disorders have been described, as well as some which are recessively inherited. The chief clinical consequence of severe hypertriglyceridaemia (types I and V) is acute pancreatitis whereas those phenotypes characterized by hypercholesterolaemia alone (type IIa) or combined with hypertriglyceridaemia (types IIb, III and IV) are associated with premature vascular disease.

PRIMARY HYPERTRIGLYCERIDAEMIA

Under this heading will be considered subjects presenting with predominant hypertriglyceridaemia (types I, IV or V phenotypes) without any obvious secondary cause. Evidence of the hereditary basis of these disorders is often only presumptive but investigation of first-degree relatives may reveal partial metabolic defects, suggestive of recessive inheritance.

Familial lipoprotein lipase deficiency

This rare disorder, known also as familial type I hyperlipoproteinaemia, is characterized by marked hypertriglyceridaemia and chylomicronaemia and usually presents in childhood. It is due to the inheritance of a recessive mutation which causes deficiency of the extrahepatic enzyme lipoprotein lipase, the rate-limiting step in chylomicron clearance. This results in a failure of lipolysis and accumulation of chylomicrons in plasma. The main clinical features are recurrent episodes of abdominal pain, often resembling acute pancreatitis, eruptive xanthomata, lipaemia retinalis and hepatosplenomegaly, associated with serum triglycerides in the region of 50–100 mmol/l.[2]

The gross chylomicronaemia results in an increase in serum cholesterol as well as triglyceride. VLDL levels are usually normal or decreased, whereas LDL and high density lipoprotein (HDL) levels are markedly reduced. Plasma lipoprotein lipase levels after an intravenous dose of heparin (10 IU/kg body weight) are less than 10% of normal. Lipoprotein lipase can be distinguished from other lipolytic enzymes by various techniques including inhibition by sodium chloride or specific antibodies, or by chromatographic separation on heparin-sepharose affinity columns.

The main aim of treatment is to minimize chylomicron formation by decreasing the intake of long-chain triglyceride to less than 50 g/day. There is no increased susceptibility to atherosclerosis in this condition, the chief complication being acute pancreatitis. The risk of this occurring is minimized if plasma triglyceride levels can be kept below 20 mmol/l.

Familial ApoC-II deficiency

This disorder is due to a recessively-inherited mutation, which in the homozygous state, results in the absence from plasma of normal apolipoprotein (Apo)C-II, the activator of lipoprotein lipase, with a consequent defect of lipolysis and hypertriglyceridaemia.[3] At least 4 distinct functionally-inactive ApoC-II variants have been described. Lipoprotein lipase is present in normal amounts but cannot hydrolyze chylomicrons or VLDL in the absence of normal ApoC-II. The diagnosis can sometimes be made by demonstrating an absent or anomalous band of ApoC-II on isoelectric focussing of delipidated VLDL.

Homozygotes have triglycerides in the range 15–107 mmol/l, usually with a type V phenotype, and frequently develop acute pancreatitis.[4] Premature vascular disease is unusual but has been described. The hypertriglyceridaemia responds dramatically to infusions of normal plasma, albeit only temporarily. Heterozygotes exhibit a 30–50% decrease in ApoC-II levels but remain normolipaemic.

Familial hepatic lipase deficiency

Two pairs of brothers have been described with this sydrome, one in Sweden, the other in Canada. Presenting features were hypertriglyceridaemia and absence of post-heparin hepatic lipase activity in plasma, whereas lipoprotein lipase activity was normal.[5] The Canadian index patient had raised levels of both serum cholesterol and triglyceride, corneal arcus, eruptive xanthomata and palmar striae, together with clinical and electrocardiographic evidence of myocardial ischaemia. The hyperlipidaemia failed to respond to clofibrate, despite the presence of β-VLDL and a type III pattern on lipoprotein electrophoresis.

The lipoprotein abnormalities in hepatic lipase deficiency reveal two characteristic features. Firstly, the HDL fraction consists of abnormally large, triglyceride-rich particles with the density of HDL_2, while HDL_3 particles are absent. This presumably reflects lack of hydrolysis of HDL_2 triglycerides, since hepatic lipase normally converts HDL_2 to HDL_3. The second lipoprotein alteration is accumulation of β-VLDL and IDL, reflecting failure of clearance of VLDL remnants. Accumulation of cholesterol-rich VLDL remnants is also seen in type III hyperlipoproteinaemia, which usually results from homozygosity for $ApoE_2$ but all the hepatic lipase-deficient patients had the normal $ApoE_3$ isoform.

Familial hypertriglyceridaemia

This disorder is usually subdivided according to whether the predominant phenotype of affected individuals is type IV or type V. However, there is overlap within families and it is probable that similar genetic abnormalities are responsible for both varieties of the disorder but with a more severe expression in those with a type V phenotype.

Type IV

Familial type IV hyperlipoproteinaemia is characterized by moderate hypertriglyceridaemia due to increased levels of VLDL. Goldstein et al.[6] studied the relatives of hypertriglyceridaemic survivors of myocardial infarction and found a bimodal distribution of fasting triglycerides consistent with autosomal dominant inheritance. They estimated the frequency of the disorder in the population at 0.2–0.3% but noted that it was expressed less frequently than expected in childhood. Mean fasting values of serum cholesterol and triglyceride in affected adults were 6.2 and 3.0 mmol/l respectively. The majority exhibited a type IV phenotype but some families also had members with a type V phenotype.

In contrast with familial combined hyperlipidaemia there were no individuals with type IIa or IIb phenotypes. The nature of the genetic defect remains to be determined but affected subjects have larger than normal VLDL particles with an increased triglyceride: ApoB ratio, accompanied by a decrease in HDL cholesterol.[7] Turnover studies show that VLDL triglyceride synthesis is increased to a greater extent than VLDL-ApoB synthesis and that the fractional catabolic rate of both VLDL components is reduced.[8,9] The latter phenomenon appears to reflect saturation of a normal clearance mechanism since post-heparin lipolytic activity (PHLA) is usually normal. Free fatty acid (FFA) flux into triglyceride is increased in type IV subjects, especially when they are placed on a high carbohydrate intake,[10] and the resultant increase in VLDL synthesis is accompanied by a decrease in the proportion of VLDL converted to LDL.[11] This maintains plasma LDL cholesterol levels within the normal range, nor is there any increase in LDL-ApoB levels, in contrast with familial combined hyperlipidaemia.

The severity of the hypertriglyceridaemia is aggravated by administration of corticosteroids or oestrogens, which can sometimes lead to acute pancreatitis. Glucose intolerance and hyperuricaemia are common accompaniments but there are no specific clinical features or biochemical markers. There are conflicting data as to whether the risk of myocardial infarction is increased[6] or is not increased[12] in this condition.

Management involves adherence to a modified fat diet designed to achieve ideal body weight, avoidance of sucrose and alcohol, and encouragement of physical activity. Drug therapy may also be required, either with nicotinic acid or one of the fibrates. However, administration of the latter sometimes results in an undesirable rise in LDL, as is also observed when type IV patients are treated with the fish oil preparation, Maxepa.[13]

Type V

This uncommon disorder has features of both type IV and type I hyperlipoproteinaemia, as would be expected in view of the increase in VLDL and chylomicrons which are its hallmark. Unlike type I it seldom presents in childhood and post-heparin lipoprotein lipase and hepatic lipase activities are usually normal. However, there is a similar liability to develop attacks of acute pancreatitis.[2] Other features are eruptive xanthomata, glucose intolerance, hyperuricaemia and peripheral neuropathy. The pancreatitis has been attributed to hydrolysis of triglyceride within the pancreas by lipase, the consequent release of FFA causing local damage to the gland. Hypertriglyceridaemia in type V subjects is accentuated by obesity and alcohol consumption. The mode of inheritance is uncertain but as mentioned previously there seems to be overlap with familial type IV hyperlipoproteinaemia. Turnover studies show similar increases in VLDL ApoB synthesis in type IV and V patients but a more marked decrease in fractional catabolic rate among the latter.[11]

Although there have been isolated case reports of coronary heart disease (CHD) in type V patients there was no evidence of undue predisposition to CHD in the 32 families studied by Greenberg et al.[14] Acute pancreatitis is the major complication and every effort should be made to avoid precipitating factors such as oestrogens and alcohol so as to maintain triglyceride levels below 20 mmol/l. Dietary control is often difficult but triglyceride synthesis can be reduced by nicotinic acid or by large doses of fish oils rich in omega-3 fatty acids.[15]

PRIMARY HYPERCHOLESTEROLAEMIA

Primary hypercholesterolaemia includes all forms of hypercholesterolaemia due to an increase in LDL or HDL for which there is no secondary cause. Familial hypercholesterolaemia (FH) has a well-established genetic basis whereas this is less clear-cut in polygenic hypercholesterolaemia and familial hyperαlipoproteinaemia. Familial combined hyperlipidaemia, which sometimes presents as hypercholesterolaemia alone, is dealt with later.

Familial hypercholesterolaemia (FH)

FH, also known as familial type II hyperlipoproteinaemia, affects approximately 0.2% of the population. Commonly this is due to inheritance of one mutant gene encoding the LDL receptor which causes heterozygous FH. Very occasionally inheritance of two mutant genes occurs, giving rise to homozygous FH. Inheritance of two dissimilar mutants results in compound heterozygotes but these cannot be distinguished clinically from homozygotes.[16]

The LDL receptor normally plays a major role in the catabolism of LDL and deficiency of LDL receptors results in accumulation

Table 1 Range of plasma or serum cholesterol values observed in FH homozygotes and heterozygotes throughout the world

Group	n	Total cholesterol (mmol/l)
Homozygotes	165	18.4–20.3
Heterozygotes	978	8.9–10.8

Data cited by Thompson et al.[21]

of LDL in plasma and hypercholesterolaemia from birth. Total cholesterol levels are roughly twice normal in adult heterozygotes and four times normal in homozygotes, as shown in Table 1. Triglyceride levels are usually normal in affected children, most of whom exhibit a type IIa phenotype, but a type IIb phenotype is quite common in adults. HDL cholesterol is often reduced, especially in homozygotes.

Homozygous FH

Clinically, homozygous FH is characterized by extreme hypercholesterolaemia and the onset in childhood of cutaneous xanthomata, typically planar or tuberose, plus tendon xanthomata and corneal arcus (Table 2). Levels of cholesterol in plasma correlate inversely with the severity of the LDL receptor deficit, which depends upon the nature of the underlying gene defect.[17] The deficit is more marked with mutations which impair the ability to produce receptors (receptor-negative) than with mutations leading to the formation of mature but abnormal receptors (receptor-defective). Turnover studies show an almost complete absence of receptor-

Table 2 Clinical characteristics of seven familial hypercholesterolaemic homozygotes studied at Hammersmith Hospital (from Ref. 18)

Patient	Sex	Age (years)	Cholesterol (mmol/l)	Triglyceride (mmol/l)	Onset of xanthomatas (years)	Aortic gradient (mmHg)	Receptor status
P.A.–S.	F	23†	26	0.5	0.5	75	Negative
N.E.	M	19†	21	–	1.5	80	Defective
R.W.	M	31†	21	0.9	5	40	Defective
D.L.	M	24	21	1.1	9	20	Defective
M.M.	M	23	19	0.9	5	0	Defective
Y.M.	M	23†	17	–	5	30	Defective
A.R.	F	36	16	2.2	12	34	Defective

†died

mediated catabolism of ApoB, with a low fractional catabolic rate of both intermediate density lipoprotein (IDL) and LDL. ApoB synthesis is twice normal, the increase being partly via a VLDL-independent pathway involving direct secretion of LDL, and partly due to increased conversion of IDL to LDL.

Atheromatous involvement of the aortic root is always evident by puberty as manifested by an aortic systolic murmur, a gradient across the aortic valve and angiographic narrowing of the aortic root together with coronary ostial stenosis.[18] Sudden death from acute coronary insufficiency before 30 was the rule before the recent introduction of better methods of treatment. At *post mortem* the aortic valve, sinuses of Valsalva and ascending arch of the aorta are grossly infiltrated with atheroma, with similar but less severe changes in all other major arteries.[19]

The chief determinant of the age of onset of CHD and the likelihood of premature death appears to be LDL receptor status.[16] Pooled data show that 60% of receptor-negative homozygotes exhibited CHD before the age of 10 years whereas this was never observed in receptor-defective patients until after that age. Furthermore, 26% of receptor-negative subjects had died from CHD before the age of 25 years compared with only 4% of receptor-defective homozygotes. Female gender does not protect against the cardiovascular complications of homozygous FH, possibly reflecting the lack in homozygotes of any sex difference in HDL cholesterol.[20]

The management of homozygous FH presents a major therapeutic challenge. Dietary and drug regimens have little impact and the safest and most effective means of reducing cholesterol levels and improving survival is to undertake plasma exchange or LDL apheresis at 2-weekly intervals.[21] Liver transplantation remedies the hepatic deficiency of LDL receptors and results in near normal cholesterol levels but necessitates long-term immunosuppression.

Heterozygous FH

Screening the children or siblings of an affected subject should lead to early detection of heterozygous FH but often it remains undiagnosed until the onset of cardiovascular symptoms in adult life. In addition to hypercholesterolaemia there are usually visible signs of cholesterol deposition, such as corneal arcus, xanthelasma and tendon xanthomata. Characteristic sites for the latter are the extensor tendons on the back of the hands and elbows, the Achilles tendons and the patellar tendon insertion into the pretibial tuberosity.

Turnover studies show a roughly 50% decrease in receptor-mediated catabolism of LDL-ApoB which results in a reduced fractional catabolic rate and prolonged half-life of LDL in plasma. LDL-ApoB synthesis is often increased but these abnormalities are less marked than in homozygotes. The cholesterol:ApoB ratio of LDL is raised, reflecting a relative increase in the proportion of light LDL present in plasma.

Diagnosis of heterozygous FH at birth is best achieved by measuring LDL cholesterol in cord blood. It has been shown that most infants with a cord blood LDL cholesterol of > 1.1 mmol/l will have an LDL cholesterol above the 95th centile after the age of 1 year but measuring total cholesterol in cord blood is unhelpful. Between the ages of 1 to 16 years serum total cholesterol levels are nearly twice as high in heterozygotes as in their unaffected siblings but the diagnosis cannot be made with confidence when the value is in the range 6.5–7.0 mmol/l, even in children with an affected parent. The high proportion of false positives in children with normal LDL levels and of false negatives in children with FH emphasizes the need to determine LDL cholesterol rather than total cholesterol when screening for FH in childhood.

Tendon xanthomata, the clinical hallmark of FH, are an age-related phenomenon. This is illustrated by an analysis of patients with definite heterozygous FH, i.e. hypercholesterolaemia plus tendon xanthomata in patient or in a first degree relative, attending 10 lipid clinics in Britain,[22] as shown in Figure 1. Overall the percentage of males and females with tendon xanthomata was very similar (75% and 72% respectively), but CHD was present in a higher proportion of males than females between the ages of 30–59 years, as shown in Figure 2.

The high frequency and premature onset of CHD in heterozygous FH has been well documented as has its much lower incidence in females as compared with males, in whom the onset of symptoms occurs 9–10 years earlier.[23] Females lose this advantage if they are smokers. It has been estimated that CHD occurs about 20 years earlier in FH than in the remainder of the population, with an accompanying decrease in life expectancy.

An inverse correlation between HDL cholesterol and CHD has been observed in FH patients of both sexes.[24] In addition lipoprotein(a) levels are much higher in FH heterozygotes with CHD than in those without, despite similar levels of LDL cholesterol.[25]

The influence of age, sex and lipid levels on the presence of tendon xanthomata and CHD is summarized in Table 3. An LDL

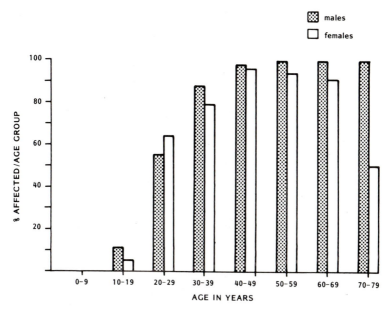

Fig. 1. Proportion of male and female heterozygotes with tendon xanthomata according to deciles of age. (Adapted by permission[22]).

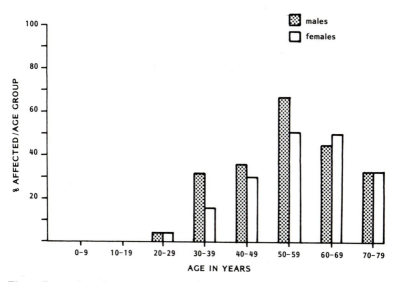

Fig. 2. Proportion of male and female heterozygotes with coronary heart disease (angina and/or history of myocardial infarction) according to deciles of age. (Adapted by permission[22]).

Table 3   Factors associated with development of tendon xanthomata and coronary heart disease in heterozygotes by age of 50 years (Published by permission from Thompson: J Inher Metab Dis 1988; 11: 18–28)

	Tendon xanthomata	Coronary heart disease
Increasing age	+ +	+ +
Male sex	−	+ +
Increased LDL-cholesterol*	+	+
Decreased HDL-cholesterol	−	+
Increased triglyceride	−	±
Tendon xanthomata		±

* > 7.8 mmol/l

cholesterol above 7.8 mmol/l seems to be a prerequisite for the development of both tendon xanthomata and CHD before the age of 50. There is a strong association between being male and developing CHD, which is presumably related to men having lower HDL cholesterol or higher triglyceride levels compared with women with FH. The absence of any relationship between HDL cholesterol and tendon xanthomata is noteworthy, implying as it does that cholesterol deposition and removal from tissues are unaffected by variations in plasma HDL.

The treatment of heterozygous FH usually involves drug therapy with an anion-exchange resin such as cholestyramine. In adult patients with a type IIb phenotype this may need supplementing with nicotinic acid or one of the fibric acid derivatives. Combination drug therapy is often necessary to achieve optimal control of hyper-cholesterolaemia and elimination of other risk factors is vital, especially smoking. Partial ileal bypass was useful in patients intolerant of resins but the introduction of hydroxy methyl glutaryl co-enzyme A (HMG CoA) reductase inhibitors has rendered this operation virtually obsolete. These drugs given either alone or together with anion-exchange resins have revolutionized the treatment of heterozygous FH, enabling reductions in LDL cholesterol of over 40% to be achieved on a long-term basis, as illustrated in Figure 3.

Familial defective ApoB$_{100}$

Using gene probes it has been recently shown that a single base substitution in the codon for arginine 3500 in the ApoB gene gives rise to a form of ApoB$_{100}$ which impairs the ability of LDL to bind to the LDL receptor.[27] Affected individuals have moderate

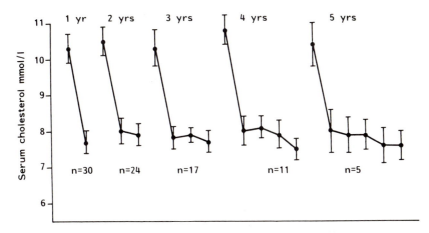

Fig. 3. Decreases in serum cholesterol observed in FH heterozygotes given lovas-tatin as an adjunct to conventional lipid-lowering therapy. In those receiving an anion-exchange resin or having undergone partial ileal bypass the combination with lovastatin resulted in a mean decrease in LDL cholesterol of 40% and a mean increase in HDL cholesterol of 23%. (Reproduced with permission from Ref. 26).

hypercholesterolaemia without any hallmarks other than a raised LDL. IDL appears to be cleared normally, in contrast to FH, presumably via binding of its ApoE to the LDL receptor. In view of the size of the gene it seems probable that other mutations affecting the receptor-binding properties of $ApoB_{100}$ may exist and represent a hitherto unrecognized cause of monogenic hyper-cholesterolaemia.

Polygenic hypercholesterolaemia

Plasma cholesterol levels are under the control of many different genes and environmental factors the summated effects of which determine the distribution of cholesterol levels in the population. Clustering in an individual of several genes which tend to induce moderate elevations of plasma cholesterol should theoretically result in polygenic hypercholesterolaemia. Indeed Goldstein et al.,[6] in their study of survivors of myocardial infarction identified a group of patients with elevated plasma cholesterol levels in which the distribution of cholesterol in the families of affected probands was unimodal but shifted towards a higher mean. This entity was defined as polygenic hypercholesterolaemia and was attributed to several independent genes clustering in any one individual. In patients with primary hypercholesterolaemia not due to FH the frequency of the ε4 allele is significantly increased. Since this allele is associated with elevated LDL-cholesterol in the general population it can be regarded as one of the genes contributing to polygenic hypercholesterolaemia.[28] Estimates of the prevalence of polygenic hypercholesterolaemia will vary according to the arbitrary definition of the upper limit of normal for serum cholesterol. Goldstein et al.[6] identified it in 14% of their hyperlipidaemic survivors of myocardial infarction, using a value of 7.4 mmol/l as their cut-off, as compared with the 10% who had FH. Obviously the lower the cut-off value used the higher will be the estimated frequency of polygenic hypercholesterolaemia relative to FH.

Familial hyperαlipoproteinaemia

Hyperαlipoproteinaemia, defined as an HDL cholesterol above the 90th percentile, sometimes occurs on a familial basis. Familial hyperαlipoproteinaemia is a heterogeneous entity in that some families show a clear-cut autosomal dominant pattern of inheritance while in others the features suggest interaction between polygenic influences and common environmental factors within the household. The increase in HDL cholesterol reflects rises in both HDL_2 and HDL_3. The syndrome tends to be associated with a decreased frequency of CHD and longevity.[29] Affected individuals require reassurance rather than treatment.

PRIMARY MIXED HYPERLIPIDAEMIAS

This heading covers a disparate group of disorders with little in common other than the presence of a mixed form of hyperlipidaemia. In some instances hypertriglyceridaemia and hypercholesterolaemia are equally prominent, as in type III hyperlipoproteinaemia, whereas in others the increase is predominantly in cholesterol, as often occurs in familial combined hyperlipidaemia.

Type III hyperlipoproteinaemia

This disorder, also known as familial dysβlipoproteinaemia,[30] is characterized by the accumulation in plasma of chylomicron and VLDL remnants which fail to get cleared at a normal rate by hepatic receptors. In vitro the LDL or ApoB,E receptor binds with high affinity particles containing either $ApoE_3$ or $ApoE_4$; the

Table 4 Some of the apolipoprotein E polymorphisms associated with type III hyperlipoproteinaemia (Adapted by permission from Wardell et al.: J Clin Invest 1987; 80: 483–490. Copyright: American Society for Clinical Investigation)

Name	Charge relative to parent E3	Molecular defect	Receptor binding activity relative to E3
E1–Harrisburg†	−2	$Lys_{146} \rightarrow Glu$	Defective
E3–A‡	0	$Cys_{112 \rightarrow 142}Arg \rightarrow Cys$	<5%
E3–Leiden	0	Tandem repeat of residues 121–127§	25%
E2	−1	$Arg_{158} \rightarrow Cys$	<2%
E2*	−1	$Arg_{145} \rightarrow Cys$	45%
E2**	−1	$Lys_{146} \rightarrow Gln$	40%
E2–Christchurch	−1	$Arg_{136} \rightarrow Ser$	41%
E–Bethesda	−2	–	–
E–Deficiency	–	–	–

†Rall et al.: J Clin Invest 1989; 83: 1095–1011
‡Mann et al.: Arteriosclerosis 1988; 8: 612a
§Wardell et al.: J Biol Chem 1989; 264: 21205–21210

latter differs from $ApoE_3$ in the substitution of arginine for cysteine at position 112 in the amino acid sequence. However, particles containing $ApoE_2$, in which there is substitution of cysteine for arginine at position 158, show virtually no binding to the ApoB,E receptor.[31] Most patients with type III hyperlipoproteinaemia are homozygous for this form of $ApoE_2$ but it has also been reported in association with the rare variants of $ApoE_2$ and $ApoE_3$ listed in Table 4, each of which exhibits reduced binding to the LDL receptor as compared with normal $ApoE_3$. The disorder has also been described in individuals with complete deficiency of ApoE.[32]

Inheritance of a defective form of ApoE is usually insufficient to give rise to clinically evident type III hyperlipoproteinaemia in that the frequency of the $ApoE_2/E_2$ phenotype in most populations is 1:100 whereas the prevalence of this form of hyperlipidaemia is only 1:5000. It seems that in addition to inheritance of a defective form of ApoE other metabolic abnormalities must also be present before hyperlipidaemia will ensue. These include obesity, diabetes, hypothyroidism or other genetic disorders such as FH and familial combined hyperlipidaemia.[33] These presumably compound the remnant clearance defect either by decreasing ApoB,E receptor expression, as occurs in FH and hypothyroidism, or by increasing VLDL synthesis and thus promoting remnant formation, as occurs in obesity and familial combined hyperlipidaemia.

Clinical features include corneal arcus, xanthelasma, tuberoeruptive xanthomata and, pathognomonically, palmar striae. Serum cholesterol and triglyceride are both elevated, usually to about 10 mmol/l. On ultracentrifugation the density (d) < 1.006 fraction contains cholesterol-rich remnants which have β-mobility on lipoprotein electrophoresis (β-VLDL). The diagnosis can be confirmed by ApoE phenotyping. LDL cholesterol is reduced due to decreased conversion of IDL to LDL[31] but despite this, atherosclerosis is common and presumably reflects the atherogenic properties of the β-VLDL particles. Vascular disease occurs in over 50% of patients, involving not only the coronary tree but also peripheral and cerebral vessels. Glucose intolerance and hyperuricaemia are also common and acute pancreatitis can sometimes occur.

Management of type III hyperlipoproteinaemia involves remedying any obvious precipitating factors, such as hypothyroidism, diabetes, obesity or iatrogenic influences. In addition, most patients will require therapy with a fibric acid derivative such as bezafibrate or gemfibrozil. Providing body weight can be controlled by diet, administration of one of these drugs often results in virtual normalization of serum lipids, rapid regression of cutaneous xanthomata and amelioration of ischaemic symptoms. Anion-exchange resins aggravate the hypertriglyceridaemia but HMG CoA reductase inhibitors are sometimes useful, especially in double heterozygotes for type III and FH.[34]

Familial combined hyperlipidaemia (FCH)

This entity was first described by Goldstein et al.[6] in hyperlipidaemic survivors of a myocardial infarction, who showed that 30% of such individuals manifested elevations of both cholesterol and triglyceride, with a variable pattern of phenotypic expression within families. Overall roughly 50% of the relatives of affected subjects were hyperlipidaemic, one third hypercholesterolaemic (type IIa), one third hypertriglyceridaemic (type IV or V) and one third with both abnormalities (type IIb). These authors concluded that the disorder was inherited in a monogenic manner but others have claimed that the pattern of transmission is more consistent with polygenic inheritance.[35] The disorder is undoubtedly familial and relatively common, occurring in up to 0.5% of the general population. The nature of the genetic defect is unknown but the disorder is characterized by increased synthesis of $ApoB_{100}$, as manifested by high rates of turnover of both VLDL and LDL-ApoB.[36]

VLDL-triglyceride synthesis is also increased but to a lesser extent than in familial hypertriglyceridaemia whereas VLDL-ApoB synthesis is increased to a more marked extent than in the latter disorder.[37] Furthermore the proportion of VLDL particles converted to LDL is normal in FCH whereas it is subnormal in familial hypertriglyceridaemia.[38] ApoB levels are raised in FCH, usually reflecting increases in LDL but sometimes reflecting increases in VLDL-ApoB.[39] VLDL particle size tends to be reduced, in contrast to familial hypertriglyceridaemia, and the concentration of IDL is increased. The LDL cholesterol:ApoB ratio and HDL cholesterol concentration both tend to be low, especially in subjects with marked hypertriglyceridaemia.

There are no distinctive clinical features in FCH and the diagnosis depends upon demonstrating multiple phenotypes within the family. In practice individuals with a type IIb phenotype who do not have tendon xanthomata are often presumed to have FCH. The condition is associated with an increased risk of atherosclerosis and it is estimated that it occurs in at least 15% of patients with CHD below the age of 60.[36]

Diet alone is often insufficient to control FCH. The main aim of therapy is to reduce excessive synthesis of VLDL and one way of achieving this is by administration of nicotinic acid. Fibric acid derivatives are easier to take but have the drawback in type IV patients of promoting conversion of VLDL to LDL and thereby increasing LDL cholesterol. This problem can be overcome by concomitant administration of either an anion-exchange resin or an HMG CoA reductase inhibitor.

HyperApobetalipoproteinaemia

The term hyperapoβlipoproteinaemia (hyperApoB) was first coined by Sniderman et al.[40] to describe a syndrome characterized by an increased concentration of LDL-ApoB (>120 mg/dl) in plasma despite a normal concentration of LDL cholesterol (<5 mmol/l). Affected individuals are often hypertriglyceridaemic, and show an apparent predisposition to atherosclerosis of the coronary, cerebral and peripheral arteries. These features are also found in FCH and there is considerable overlap between these two entities. An increase in VLDL-ApoB synthesis is seen in both disorders, which leads to an increased rate of synthesis and thus an expanded pool of LDL-ApoB, despite LDL being catabolized at a normal rate.[41] Since hyperApoB exhibits similar phenotypic features to FCH and both disorders seem to have the same metabolic defect, namely overproduction of ApoB, this suggests that they may be identical. However, until such time as the genetic defects have been identified it is a moot point whether primary hyperApoB represents a subgroup of FCH patients with type IV phenotypes or a genetically distinct disorder.

REFERENCES

1 Beaumont JL, Carlson LA, Cooper GR, Fejfar Z, Fredrickson DS, Strasser T. Classification of hyperlipidaemias and hyperlipoproteinaemias. Bull WHO 1970; 43: 891–908
2 Nikkila EA. Familial lipoprotein lipase deficiency and related disorders of chylomicron metabolism. In: Stanbury JB, Wyngaarden JB, Fredrickson DS, Goldstein JL, Brown MS, eds. Metabolic Basis of Inherited Disease. 5th edn. New York: McGraw Hill, 1983: pp 622–642
3 Breckenridge WC, Little A, Steiner G, Chow A, Poapst M. Hypertriglyceridemia associated with deficiency of apolipoprotein C-II. N Engl J Med 1978; 298: 1265–1273
4 Cox DW, Breckenridge WC, Little JA. Inheritance of apolipoprotein C-II deficiency with hypertriglyceridemia and pancreatitis. N Engl J Med 1978; 299: 1421–1427
5 Breckenridge WC, Little JA, Alaupovic P et al. Lipoprotein abnormalities associated with a familial deficiency of hepatic lipase. Atherosclerosis 1982; 45: 161–179
6 Goldstein JL, Schrott HG, Hazzard WR, Bierman EL, Motulsky AG. Hyperlipidemia in coronary heart disease. II. Genetic analysis of lipid levels in 176 families and delineation of a new inherited disorder, combined hyperlipidemia. J Clin Invest 1973; 52: 1544–1568
7 Brunzell JD, Albers JJ, Chait A, Grundy SM, Groszek E, McDonald GB. Plasma lipoproteins in familial combined hyperlipidemia and monogenic familial hypertriglyceridemia. J Lipid Res 1983; 24: 147–155
8 Chait A, Alberts JJ, Brunzell JD. Very low density lipoprotein overproduction in genetic forms of hypertriglyceridaemia. Eur J Clin Invest 1980; 10: 17–22
9 Kissebah AH, Alfarsi S, Adams PW. Integrated regulation of very low density lipoprotein triglyceride and apolipoprotein B kinetics in man: normolipemic subjects, familial hypertriglyceridemia and familial combined hyperlipidemia. Metabolism 1981; 30: 856–868
10 Quarfordt SH, Frank A, Shames DM, Berman M, Steinberg D. Very low density lipoprotein triglyceride transport in type IV hyperlipoproteinaemia and the effects of carbohydrate-rich diets. J Clin Invest 1970; 49: 2281–2297
11 Packard CJ, Shepherd J, Joerns S, Gotto AM, Taunton OD. Apolipoprotein B metabolism in normal, type IV and type V hyperlipoproteinemic subjects. Metabolism 1980; 29: 213–222
12 Brunzell JD, Schrott HG, Motulsky AG, Bierman EL. Myocardial infarction in the familial forms of hypertriglyceridemia. Metabolism 1976; 25: 313–320
13 Sullivan DR, Sanders TAB, Trayner IM, Thompson GR. Paradoxical elevation of LDL apoprotein B levels in hypertriglyceridaemic patients and normal subjects ingesting fish oil. Atherosclerosis 1986; 61: 129–134
14 Greenberg BH, Blackwelder WC, Levy RI. Primary type V hyperlipoproteinemia. A descriptive study in 32 families. Ann Intern Med 1977; 87: 526–534
15 Phillipson BE, Rothrock DW, Connor WE, Harris WS, Illingworth DR.

Reduction of plasma lipids, lipoproteins, and apoproteins by dietary fish oils in patients with hypertriglyceridemia. N Engl J Med 1985; 312: 1210–1216
16 Goldstein JL, Brown MS. Familial hypercholesterolemia. In: Stanbury JB, Wyngaarden JB, Fredrickson DS, Goldstein JL, Brown MS, eds. The Metabolic Basis of Inherited Disease. 5th edn. New York: McGraw-Hill, 1983; pp 672–712
17 Sprecher DL, Hoeg JM, Schaefer EJ et al. The association of LDL receptor activity, LDL cholesterol level, and clinical course in homozygous familial hypercholesterolaemia. Metabolism 1985; 34: 294–299
18 Allen JM, Thompson GR, Myant NB, Steiner R, Oakley CM. Cardiovascular complications of homozygous familial hypercholesterolaemia. Br Heart J 1980; 44: 361–368
19 Buja LM, Kovanen PT, Bilheimer DW. Cellular pathology of homozygous familial hypercholesterolemia. Am J Pathol 1979; 97: 327–345
20 Seftel HC, Baker SG, Sandler MP et al. A host of hypercholesterolaemic homozygotes in South Africa. Br Med J 1980; 281: 633–636
21 Thompson GR, Barbir M, Okabayashi K, Trayner I, Larkin S. Plasmapheresis in familial hypercholesterolemia. Arteriosclerosis 1989; 9 (Suppl I): I-152–I-157
22 Thompson GR, Seed M, Niththyananthan, McCarthy S, Thorogood M. Genotypic and phenotypic variation in familial hypercholesterolaemia. Arteriosclerosis 1989; 9 (Suppl I): I-75–I-80
23 Slack J. Risk of ischaemic heart disease in familial hyperlipoproteinaemic states. Lancet 1969; ii: 1380–1382
24 Moorjani S, Gagné C, Lupien P-J, Brun D. Plasma triglycerides related decrease in high density lipoprotein cholesterol and its association with myocardial infarction in heterozygous familial hypercholesterolemia. Metabolism 1986; 35: 311–316
25 Seed M, Hoppichler F, Reaveley D et al. Relation of serum lipoprotein(a) concentration and apolipoprotein(a) phenotype to coronary heart disease in patients with familial hypercholesterolemia. N Engl J Med 1990; 322: 1494–1499
26 Mater VMG, Thompson GR. HMG CoA reductase inhibitors as lipid-lowering agents: five years experience with lovastatin and an appraisal of simvastatin and pravastatin. Q J Med 1990; 74: 165–175
27 McCarthy BJ, Soria L, Ludwig EM et al. An arginine 3500→glutamine mutation in familial defective apoB-100 subjects with LDL defective in binding to the apoB₁E (LDL) receptor. Circulation 1988; 78 (suppl II): II66
28 Utermann G. Apolipoprotein polymorphism and multifactorial hyperlipidaemia. J Inher Metab Dis 1988; 11 (suppl I): 74–86
29 Glueck CJ, Fallat RW, Millett F, Gartside P, Elston RC, Go RCP. Familial hyper-alpha-lipoproteinemia: studies in eighteen kindreds. Metabolism 1975: 24: 1243–1265
30 Havel RJ. Familial dysbetalipoproteinemia. New aspects of pathogenesis and diagnosis. Med Clin N Am 1982; 66: 441–454
31 Mahley RW, Innerarity TL, Rall SC, Weisgraber KH. Plasma lipoproteins: apolipoprotein structure and function. J Lipid Res 1984; 25: 1277–1294
32 Schaefer EJ, Gregg RE, Ghiselli G et al. Familial apolipoprotein E deficiency. J Clin Invest 1986; 78: 1206–1219
33 Brown MS, Goldstein JL, Fredrickson DS. Familial type 3 hyperlipoproteinemia (dysbetalipoproteinemia). In: Stanbury JB, Wyngaarden JB, Fredrickson DS, Goldstein JL, Brown MS, eds. The Metabolic Basis of Inherited Disease. 5th edn. New York: McGraw Hill, 1983: pp 655–671
34 Thompson GR, Ford J, Jenkinson M, Trayner I. Efficacy of mevinolin as adjuvant therapy for refractory familial hypercholesterolaemia. Q J Med 1986; 60: 801–809

1004   LIPIDS AND CARDIOVASCULAR DISEASE

35 Nikkila EA, Aro A. Family study of serum lipids and lipoproteins in coronary heart disease. Lancet 1973, i: 954–959
36 Grundy SM, Chait A., Brunzell JD. Familial combined hyperlipidaemia workshop. Arteriosclerosis 1987; 7: 203–207
37 Chait A, Albers JJ, Brunzell JD. Very low density lipoprotein overproduction in genetic forms of hypertriglyceridaemia. Eur J Clin Invest 1980; 10: 17–22
38 Kissebah AH, Alfarsi S, Adams PW. Integrated regulation of very low density lipoprotein triglyceride and apolipoprotein B kinetics in man: normolipemic subjects, familial hypertriglyceridemia and familial combined hyperlipidemia. Metabolism 1981; 30: 856–868
39 Brunzell JD, Albers JJ, Chait A, Grundy SM, Groszek E, McDonald GB. Plasma lipoproteins in familial combined hyperlipidemia and monogenic familial hypertriglyceridemia. J Lipid Res 1983; 24: 147–155
40 Sniderman A, Shapiro S, Marpole D, Skinner B, Teng B, Kwiterovich PO. Association of coronary atherosclerosis with hyperapobetalipoproteinemia [increased protein but normal cholesterol levels in human low density (t) lipoproteins]. Proc Natl Acad Sci USA 1980; 77: 604–608
41 Teng B, Sniderman AD, Soutar AK, Thompson GR. Metabolic basis of hyperapobetalipoproteinemia. Turnover of apolipoprotein B in low density lipoprotein and its precursors and subfractions compared with normal and familial hypercholesterolemia. J Clin Invest 1986; 77: 663–672

British Medical Bulletin (1990) Vol. 46, No. 4, pp. 1005–1024
© The British Council 1990

Secondary hyperlipidaemia

P N Durrington
University of Manchester Department of Medicine, Manchester Royal Infirmary, Manchester, UK

Secondary hyperlipidaemia is common and occurs frequently in disorders such as obesity, alcoholism, diabetes mellitus, hypothyroidism, liver and renal diseases and as a side-effect of drug therapy, particularly for hypertension. Its management may be important to prevent complications such as coronary heart disease and acute pancreatitis. Its study provides many fascinating insights into lipoprotein pathophysiology.

Hyperlipidaemia frequently coexists with other diseases. These other diseases may occur as a complication of the hyperlipidaemia, as is the case for example when familial hypercholesterolaemia is associated with ischaemic heart disease or marked hypertriglyceridaemia with acute pancreatitis. Sometimes, however, another primary disease affects lipoprotein metabolism in such a way as to increase serum lipid concentrations. The resulting hyperlipidaemia is then properly regarded as a secondary hyperlipidaemia. Hyperlipidaemia may also be associated with another disorder, when neither occurs as the complication of the other. The best example of this would be hypertriglyceridaemia and gout which frequently coexist, but neither is the cause of the other and treatment of one does not usually influence the other. The hyperlipidaemia of gout is usually therefore wrongly classified as a secondary hyperlipidaemia, but will nonetheless be considered in this chapter as is customary.

The secondary hyperlipidaemias are important because:

(i)   the primary disease presenting as hyperlipidaemia may be an important diagnosis in its own right

(ii)   the secondary hyperlipidaemia may be a cause of morbidity, e.g. in diabetes[1] and renal disease[2] ischaemic heart disease is

0007–1420/90/0046–1005/$10.00

now the major cause of premature mortality and a major reason for this is abnormal lipid metabolism

(iii) disordered lipoprotein metabolism may accelerate the progress of the primary disease as has been suggested in renal and liver disease.[3,14] Lipoproteins probably have multiple functions besides that of lipid transport and their involvement in cell membrane homeostasis[4] and immunoregulation[5] may be critical.

The impact of secondary hyperlipidaemias depends on the milieu in which they occur and will be more extreme in people already genetically or nutritionally predisposed to hyperlipoproteinaemia. Patients with pre-existing familial hypertriglyceridaemia may, for example, develop extremely high levels of triglycerides if diabetes develops or β-adrenoreceptor blocking drugs are administered.[6] On the other hand diabetic hyperlipidaemia and ischaemic heart disease are unusual in populations such as the Japanese, who are not nutritionally disposed to hyperlipidaemia, in comparison to populations with a Northern European-style diet.[7]

The secondary hyperlipidaemias are often associated not simply with increases in levels of circulating lipoproteins, but also with changes in their composition, altering both their chemical and physical properties. This may influence the accuracy of laboratory methods developed for quantitating lipoproteins in primary hyperlipoteinaemias. It may also mean that even when lipid levels are not clearly elevated there may be qualitative changes in lipoproteins, perhaps rendering them more atherogenic, which we are only beginning to appreciate. Dyslipidaemia may thus eventually assume more importance than hyperlipidaemia in our understanding of atheroma partly as a result of the study of secondary lipoprotein disorders.

The secondary hyperlipidaemias are listed in Table 1. Some of the more important and more common of these will be discussed in the remainder of this chapter. Their major effects on serum lipoprotein concentrations are shown in Table 2. There are two early reviews of the secondary hyperlipidaemias[8,9] and a recent detailed review.[6]

DIABETES MELLITUS

For too long diabetologists have been guilty of regarding diabetes as simply a disorder of carbohydrate metabolism and insulin as a

Table 1   Diseases and physiological and pharmacological perturbations associated with secondary hyperlipidaemia

Endocrine	Diabetes mellitus Thyroid disease Pituitary disease Pregnancy
Nutritional	Obesity Alcohol Anorexia nervosa
Renal disease	Nephrotic syndrome Chronic renal failure
Drugs	β-adrenoreceptor blockers Thiazide duretics Steroid hormones Microsomal enzyme inducing agents Retinoic acid derivatives
Hepatic disease	Cholestasis Hepatocellular disorders Cholelithiasis Hepatoma Porphyria
Immunoglobulin excess	Myeloma Macroglobulinaemia Systemic lupus erythematosus
Hyperuricaemia	
Miscellaneous	Stress Intestinal malabsorption Glycogen storage disease Lipodystrophy Idiopathic hypercalcaemia of infants Hypervitaminosis D Osteogenesis imperfecta Sphingolipodystrophies Progeria Werner's syndrome Cholesteryl ester storage disease Carnitine acyl transferase deficiency Tangier disease Familial LCAT deficiency

hormone with the sole purpose of maintaining euglycaemia. Slowly there is a renaissance of the idea that diabetes is a disorder not only of carbohydrate metabolism, but also (and in many patients more importantly) of lipid and protein metabolism. Two major complications of diabetes, atherosclerosis and ketoacidosis are disorders of lipid metabolism and the possibility that there are others too must never again be neglected.

Table 2 Effect on serum lipoproteins of some secondary hyperlipidaemias

Cause	VLDL	LDL	HDL
NIDDM*	↑↑	↑	↓
IDDM*	↑	Nor↓	Nor↑
Hypothyroidism	↑	↑↑	↑
Pregnancy	↑	↑	↑
Obesity	↑	Nor↑	↓
Alcohol	↑	Nor↑	↑
Nephrotic syndrome	↑	↑↑	Nor↓
Chronic renal failure	↑	–	↓
Cholestasis	–	↑↑(LpX)	↓
Hepatocellular disease	↑(IDL)	–	↓
Hyperuricaemia	↑	–	↓

*Non-insulin-dependent diabetes (Type 2 diabetes)
**Insulin-dependent diabetes (Type 1 diabetes)

The dominant hyperlipidaemia in diabetes mellitus is hypertriglyceridaemia.[9,10] The enzyme, lipoprotein lipase, is activated by insulin.[9,10] This enzyme is located on the capillary endothelium of tissues such as adipose tissue and skeletal muscle, which are active in the catabolism of triglyceride-rich lipoproteins. Thus insulin deficiency and/or resistance associated with uncontrolled diabetes may occasionally lead to spectacular elevations of serum triglycerides sometimes to levels of 100 mmol/l or more and to the development of eruptive xathomata and occasionally other features of the chylomicronaemia syndrome.[11] Lipaemia retinalis will interfere with laser photocoagulation therapy for diabetic retinopathy, which may advance rapidly in such patients. Patients who develop this syndrome are probably genetically predisposed to hypertriglyceridaemia: there is generally a preexisting partial defect in triglyceride catabolism. Sometimes tuberoeruptive xanthomata and striate palmar xanthomata indicate that florid Type III hyperlipoproteinaemia has occurred in a genetically susceptible individual (usually an apoliprotein E_2 homozygote).

In most patients, however, whose diabetes is under reasonable glycaemic control, any persisting hypertriglyceridaemia, is not due to a major defect in triglyceride catabolism, but to overproduction of VLDL by the liver.[9,10] Nonesterified fatty acids (NEFA) arriving at the liver from the adipose tissue and skeletal muscle in increased quantities are likely to be a major reason for increased hepatic triglyceride synthesis. The release of these newly synthesized triglycerides from the liver as VLDL is likely to be facilitated by decreased insulin secretion and/or insulin resistance, which will decrease the direct inhibitory effect of insulin on the secretion of VLDL by hepatocytes.[12,14] Even in insulin-treated Type 1 diabetes the liver is likely to remain deficient in insulin since insulin administered via the subcutaneous route arrives at the liver by the systemic circulation rather than the portal vein, in which its physiological concentration is several times that in the systemic circulation.[15]

Whether insulin has a direct effect on hepatic triglyceride synthesis is undecided. The hypothesis devised by Reaven and coworkers that insulin stimulates triglyceride synthesis and that 'hyperinsulinaemia' in diabetes leads to hypertriglyceridaemia,[16] although widely quoted, fails to explain too many experimental observations for it to continue to be entertained.[6] So enshrined in dogma has it been that clinicians are often surprised to learn that there is no circumstance in man or intact animals in which the administration of insulin does not lower the serum triglyceride level.[1,6,17]

Patients with Type 1 diabetes tend to be less likely to have hypertriglyceridaemia than those with Type 2 diabetes.[6] This may be partly because insulin therapy is the rule in Type 1 diabetes and also because in Type 2 diabetes other factors predisposing to hypertriglyceridaemia such as obesity, and β-blocker and diuretic therapy are more common. There may, however, in addition be a more fundamental reason. NEFA are disposed of by the liver in three major processes: complete oxidation, partial oxidation (ketogenesis) or esterification (triglyceride synthesis). When hepatic energy requirements are met only ketogenesis or esterification are possible. The essential metabolic difference between Type 1 and Type 2 diabetes is that in Type 2 NEFA are not readily converted to ketone bodies whereas in Type 1 their entry into the mitochondria where β-oxidation occurs is facilitated for reasons which are not entirely understood.[18] Resistance to ketogenesis in Type 2 diabetes may thus carry with it a predisposition to hypertriglyceridaemia.

Hypertriglyceridaemia is a much stronger marker for coronary heart disease in diabetes than in non-diabetic populations. It is not possible to regard it as an independent risk factor without qualification, because HDL, low levels of which frequently coexist with hypertriglyceridaemia, was not measured in the WHO study.[19] It seems likely, however, that there is a definite link between hypertriglyceridaemia and coronary artery disease in diabetes. This is unlikely to involve the triglyceride-rich lipoprotein responsible for the elevated serum triglyceride levels directly, but is likely to result from the IDL produced by their catabolism[20,21] or from the small LDL particles which are apoliprotein B rich and relatively cholesterol depleted. These are associated with hypertriglyceridaemia in some patients. Whilst in Type 2 diabetics with hypertriglyceridaemia there is evidence that they may produce more small LDL,[22] in insulin-treated Type 1 patients the overwhelming effect is a cholesterol-rich LDL.[20] In both types of diabetes, however, remnant particles probably persist at higher concentration than in non-diabetics.[6,20,23] The explanation for this is uncertain. Apolipoprotein E_2 homozygosity although more prevalent in diabetes than in the general population is not sufficiently common to be the explanation.[24] Glycosylation of apolipoprotein E interfering with uptake of chylomicron remnants and IDL by the hepatic remnant receptor may, however, be important.[25] So also may be a defect in the regulation of postprandial lipoprotein metabolism in diabetes.[6] Normally the insulin which is secreted as the products of digestion enter the circulation may function to inhibit hepatic VLDL secretion and promote hepatic triglyceride storage[12] at a time when triglyceride-rich lipoprotein production is high. This would be expected to relieve pressure on triglyceride catabolic pathways, such as lipoprotein lipase and the hepatic remnant (apolipoprotein E) receptor, preventing the accumulation of remnants and IDL in the circulation postprandially. Later when insulin levels decline hepatic VLDL secretion probably increases and stored triglycerides are mobilized. In diabetes failure to suppress VLDL secretion following meals might lead to high levels of remnant particles and IDL.

Serum LDL cholesterol and apolipoprotein B levels are on average somewhat increased in Type 2 diabetes compared to a nondiabetic population whereas in Type 1 diabetes they are frequently normal or even a little lower.[6] The LDL receptor is 'up-regulated' by insulin[26] and this may partly account for the difference since the typically insulin-resistant Type 2 patient may have decreased LDL catabolism whereas the Type 1 patient may have insulin levels which in the systemic circulation are supraphysiological.[6] Glycosylation of apolipoprotein B also interferes with LDL receptor mediated uptake, at least in vitro, and might also contribute to decreased LDL catabolism.[25] More important than the absolute levels of LDL in diabetes is whether it is rendered more atherogenic, for example more cytotoxic or more readily taken up by macrophage acetyl-LDL receptors. Recent evidence suggests that apolipoprotein B in addition to undergoing glycosylation may be more subject to oxidative modification in diabetes.[27] Collagen advanced glycosylation end-products may also trap LDL in the arterial wall.[28]

Serum HDL cholesterol concentrations tend to be low in Type 2 diabetes whereas in Type 1 diabetes they are normal or even raised.[6,29] The low levels in Type 2 diabetes are largely explicable on the basis of associated hypertriglyceridaemia, obesity, cigarette-smoking, abstemiousness from alcohol[29] and drugs such as β-adrenoreceptor blockers.[30] In Type 1 diabetes these factors are less common and some other influence tends to increase HDL. Probably this is insulin therapy. A possible mechanism for this is that insulin increases HDL as a result of stimulating lipoprotein lipase activity,[31] but this remains uncorroburated.[32] The contrary suggestion that the low HDL cholesterol in Type 2 diabetes is due to preserved endogenous insulin secretion is not supported by detailed analysis.[33]

Diabetic therapy besides insulin also influences serum lipids. The carbohydrate-restricted diet which regrettably still persists in some quarters was an unfortunate diverticulum in diabetic management from which we are only beginning to emerge. Not only has it been known since the 1930's that glucose tolerance deteriorates further when such a diet is given,[34] but also, because many patients asked to restrict their carbohydrate intake chose to maintain their energy intake by eating more fat, which exacerbated their hyperlipidaemia. Oral hypoglycaemic agents in therapeutic trials do not adversely affect lipoprotein levels and improvements in glycaemia control may even produce some improvement in lipid profiles.[6] Metformin and guar gum also probably have an independent lipid-lowering action.[35] Sadly the use of sulphonylurea drugs in clinical practice is linked with decreased HDL and probably other adverse effects such as increases in triglycerides and cholesterol.[6] This is because the use of these drugs in practice is all too often associated with body weight gain. Insulin too, although its action *per se* is to lower triglycerides and cholesterol and to raise HDL, may stimulate weight gain and thus increase insulin resistance, which will tend to nullify or even reverse any beneficial effects.[36]

Proteinuria,[6,37,38] hypertension,[6,37,38] and hyperfibrinogenaemia[38,39] frequently coexist with hyperlidipaemia in diabetes and considerably increase coronary risk. Nephropathy as in the case of primary renal disease may influence lipoprotein metabolism (see later), but the major significance of proteinuria in diabetes may be that it indicates a generalised increase in vascular permeability and thus macromolecules such as LDL may enter the arterial subintima at increased rates, an effect aided and abetted by hypertension.

THYROID DISEASE

In hypothyroidism serum LDL cholesterol and less frequently serum triglycerides are raised.[6] HDL, particularly HDL_2, levels tend to be increased. There is decreased receptor-mediated LDL catabolism[40,41] and triglyceride catabolism and lipoprotein lipase activity may also be decreased.[43] Often these effects are reversible with thyroxine replacement therapy, which also restores to normal biliary cholesterol excretion, which was previously depressed. Subclinical hypothyroidism (raised serum TSH, but thyroxine in the normal range) probably does slightly influence serum LDL.[43] In one survey raised serum TSH was detected in 20% of women over the age of 40 years whose serum cholesterol exceeded 8 mmol/l.[44] Although only 5% were actually hypothyroid, findings such as these underline the importance of adequately excluding hypothyroidism in people with hypercholesterolaemia beyond mid-life.

There is a tendency for decreased LDL and HDL cholesterol in hyperthyroidism.[6] It is uncertain whether the β-migrating HDL in one report[45] represents an increase in pre-β-HDL (free apolipoprotein AI).

OBESITY

Obesity will exacerbate any primary hyperlipoproteinaemia. Its dominant effect *per se* is to produce hypertriglyceridaemia usually Type IV hyperlipoproteinaemia (occasionally Type V), but in susceptible individuals hypercholesterolaemia due to increased LDL will be exacerbated (Type IIb).[6] Android (male pattern) obesity is more likely to provoke hypertriglyceridaemia than gynoid (female pattern) obesity assessed by determining the ratio of the hips to the waist.[6] The cause of the hypertriglyceridaemia is increased hepatic VLDL production.[46] This probably results from an increased release of NEFA from adipose tissue which in turn stimulates hepatic triglyceride production.[47] The increased VLDL scretion is often matched by an enhanced catabolism due to increased lipoprotein lipase activity so that hypertriglyceridaemia does not invariably ensue.[46] It is more likely if there is a preexisting defect in triglyceride catabolism. There is increased cholesterol synthesis in obesity and it is perhaps surprising that hypercholesterolaemia is not a more common association of obesity in epidemological studies.

Serum HDL cholesterol tends to be decreased in obesity.[48] The explanation for this is uncertain since in many patients triglyceride catabolism is enhanced. Furthermore whereas serum triglycerides and cholesterol decrease during weight reduction no increase in serum HDL occurs. Sometimes, if substantial weight loss is achieved and maintained, HDL does rise. However, patients able to achieve this must be a highly selected group and it is impossible to dissociate their weight loss from other changes in the life style of the new slim individual, which might increase the serum HDL concentration.

ALCOHOL

Alcoholic beverages particularly beer and wine are energy rich and may in heavy drinkers be the cause of obesity. In addition alcohol itself affects lipoprotein metabolism. Its dominant effect is to produce hypertriglyceridaemia by increasing hepatic triglyceride synthesis.[6] There is an increased hepatic VLDL secretion. Fatty liver ensues, if this fails to keep pace with production of triglyceride. Usually Type IV hyperlipoproteinaemia is produced, but in individuals with a constitutional tendency to delayed triglyceride catabolism[49] a spectacular Type V hyperlipoproteinaemia may occur, which may be one explanation for the association between alcohol consumption and acute pancreatitis.[50] The increase in hepatic triglyceride synthesis stems partly from the ethanol-induced inhibition of oxidation of substrates other than itself.[9] This tends to divert NEFA away from oxidative pathways into triglyceride synthesis. Triglyceride synthesis is further accelerated by the increased release of NEFA from adipose tissue, particularly when ethanol is taken during fasting, or by the food-induced fatty acidaemia when alcohol is taken during a meal.[6]

Serum LDL cholesterol levels tend to be low in chronic alcoholics and HDL cholesterol to be raised, unless liver disease has developed. The effect on HDL is evident in moderate drinkers and in them is due predominantly to an effect on HDL_3, whereas in heavy drinkers the quantitatively greater increase in HDL is due to HDL_2.[48]

The recognition of occult alcoholism is obviously important in the Lipid Clinic, particularly in hypertriglyceridaemic patients prone to pancreatitis. It is not realized frequently enough that measuring serum γ-glutamyl transpeptidase is not always helpful since it may be raised in patients with hypertriglyceridaemia unrelated to alcohol.[51]

RENAL DISEASE

Nephrotic syndrome

When proteinuria occurs in patients with relatively normal creatinine clearance the predominant effect is to produce hypercholesterolaemia due to an increase in LDL.[6] The severity of the hypercholesterolaemia is often proportional to the decrease in serum albumin.[6] In man hypertriglyceridaemia is unusual in nephrotic syndrome in the absence of hypercholesterolaemia and is more likely when chronic renal failure is also present. Then it is frequently associated with decreased lipoprotein lipase activity.[52] In the rat, nephrotic syndrome causes hypertriglyceridaemia and this is the cause of some confusion in understanding the mechanism of hypercholesterolaemia in man. VLDL secretion by cultured rat hepatocytes decreases in response to albumin in the culture medium. The effect may be due to osmotic pressure or to viscosity since other macromolecules affect VLDL secretion in a similar way.[53] In man the hypercholesterolaemia is unlikely to be explained by increased production of LDL from VLDL, the serum levels of which are not raised. The intravenous infusion of albumin or other macromolecules does, however, reduce LDL levels[54] which might suggest that direct hepatic secretion of LDL without a VLDL precursor may be important.

Our research group has recently found that apolipoprotein B secretion by human hepatoma cells is suppressed by albumin. Also work by Dr C Short in our laboratory has revealed that patients with nephrotic syndrome frequently have high serum concentrations of Lp(a), which is a subfraction of LDL, which is believed to lack a VLDL precursor. Interestingly one other clinical syndrome associated with high serum Lp(a) levels is familial hypercholesterolaemia.[55] Direct evidence that the increase in Lp(a) is due to the LDL receptor defect in this condition is not yet forthcoming and it may be that direct hepatic secretion of LDL without a VLDL precursor, which is known to occur in familial hypercholesterolaemia,[56] is, as in nephrotic syndrome, responsible for the high Lp(a) levels.

Serum HDL cholesterol and apolipoprotein AI levels are usually normal or decreased in nephrotic syndrome. Even when total HDL is normal there is a shift towards particles of smaller size so that there is a decrease in the HDL_2 subfraction whilst HDL_3 often rises.[57] Dr R Neary in our laboratory has also recently shown that the even smaller pre-β-HDL (free apolipoprotein AI) levels are also increased. Apolipoprotein AI production is increased so that increased catabolism must explain the normal or low circulating levels. The increased loss of AI from the circulation is due to leakage from the kidney, and is related to the selectivity and the extent of the glomerular leak.[57] Immunoreactive apolipoprotein AI in quantities equal to the normal daily apolipoprotein AI production may be found in the urine. The AI-containing particles in the urine were recently found by Neary using gel chromatography to be predominantly in the pre-β-HDL and HDL_3 range. We believe that circulating smaller HDL particles are increased due to increased production, but that they are lost into the urine before they acquire sufficient cholesterol to be converted to HDL_2, accounting for its low levels.

Chronic renal failure without proteinaemia

Serum triglycerides are raised in renal failure both in VLDL and LDL.[6,52] There is also a tendency for remnant particles to persist in the circulation.[58] The underlying cause is uncertain, but may relate to decreased activity of both lipoprotein lipase and hepatic lipase.[52] The insulin resistance associated with renal failure does not appear to increase NEFA flux as it often does under other conditions. Haemodialysis further exacerbates hypertriglyceridaemia: the frequent use of heparin depletes lipoprotein lipase and in addition there is loss from the circulation of apolipoprotein CII, the activator of lipoprotein lipase. Chronic ambulatory peritoneal dialysis leads to absorption from the peritoneum of considerable amounts of glucose producing obesity and exacerbating hypertriglyceridaemia. In addition even when LDL cholesterol levels are not raised, LDL apolipoprotein B often is.[59]

In patients with chronic renal failure serum HDL cholesterol levels are low. The apolipoprotein AI level, however, is not decreased to the same extent due to an increase in pre-β-HDL (free apolipoprotein AI).[60] This may accumulate because of decreased LCAT activity,[52] since pre-β-HDL appears to be the preferred substrate for LCAT, which converts it to HDL_3.[61]

Following renal transplantation many of the lipoprotein abnormalities resolve if good renal function is established, but in about one quarter of patients hyperlipidaemia persists perhaps because of corticosteroid therapy, weight gain, antihypertensive therapy and possibly cyclosporin treatment.

DRUGS

A large number of drugs in common use affect serum lipoprotein concentrations[6] (Table 3). Those most commonly encountered in the 'Lipid Clinic' are diuretics and β-adrenoreceptor blocking drugs. Thiazide diuretics and probably also loop diuretics raise VLDL and LDL by mechanisms which have not been elucidated.[62] Generally the effect is small but it may be more substantial in diabetes, which they also exacerbate. Diuretics do not alter HDL levels. Beta-adrenoreceptor blockers, regardless of whether they are cardioselective or not tend to increase serum triglyceride

Table 3 Drugs affecting lipoprotein metabolism (*see* Reference 6)

Drug	VLDL	LDL	HDL
β-Adrenoreceptor blockers without ISA*	↑	–	↓
Thiazides	↑	↑	–
Oestrogens	↑	– or ↓ **	↑
Progestogens	–	↑	↓
Androgens	↓	↑	↓
Glucocorticords	– or ↑	↑	↑
Hepatic microsomal enzyme inducing agents e.g. phenytoin phenobarbitone, rifampicin, griseofulvin	–***	–***	↑
Retinoic acid derivatives e.g. etretinate	↑	–	–

*ISA = intrinsic sympathomimetic activity
**Decrease LDL in postmenopausal women
***May be unsustained increase

concentrations by an effect on VLDL and to decrease HDL cholesterol.[63,64] There is no convincing evidence that they affect total cholesterol or LDL cholesterol. Their effect on serum triglycerides may be marked in patients with pre-existing hypertriglyceridaemia[65] in whom a decrease in the clearance of triglyceride-rich lipoproteins appears to be the mechanism, perhaps because of a direct effect in reducing the activity of the enzyme, lipoprotein lipase or by diversion of blood flow away from the vascular bed of muscle which is one site where the enzyme is located. The explanation for the smaller increase in serum triglycerides in normolipidaemic people is unclear. Beta-blockers with intrinsic sympathomimetic activity (ISA) have little or no effect on serum HDL and triglycerides.[63,65] Of this class pindolol has the most ISA, but has found little favour as an antihypertensive and is unsuitable for the management of angina. Acebutol and oxprenolol, which have about half as much ISA as pindolol, but about double that of other β-blockers, may be valuable in some patients with hypertriglyceridaemia when β-blocker therapy cannot be avoided. Lobetalol, which combines α and β-blocking activity, also is reported to have little effect on serum lipoproteins.[63,64] Many reports suggest that α-blockers, calcium antagonist vasodilators, direct acting vasodilators and angiotensin-converting enzyme inhibitors are either without effect on serum lipoproteins or may even have apparently favourable effects, such as a raising HDL cholesterol.[63] There is, however, no evidence as yet that pharmacologically induced changes of this type actually do significantly reduce disease morbidity or mortality.

Oestrogens tend to raise the serum triglyceride level due to increased hepatic VLDL production.[6] Occasionally in women with pre-existing hypertriglyceridaemia, their inadvertent administration has led to gross hyperchylomcronaemia and consequently to acute pancreatitis. In most women the increase in triglycerides is small and paradoxically in Type III hyperlipoproteinaemia, improvement has been reported, possibly because oestrogen–induction of remnant receptors outweighs any deleterious effect of increased VLDL production.

Oestrogens also raise serum HDL concentrations and in postmenopausal women decrease serum LDL levels.[6] The effect of androgens is generally the converse: a decrease in serum HDL cholesterol and VLDL and an increase in LDL.[6] Progestogens raise LDL and decrease HDL to an extent which depends on their androgenicity.[6] Thus in terms of HDL and LDL change, androgens would appear to have a particularly unfavourable effect on serum lipoproteins and progestogens to have similar, although usually less marked effects, whereas oestrogens might be regarded as potentially beneficial. This latter conclusion must, however, be tempered with caution. Firstly because the opportunities to administer oestrogen preparations alone are few. Generally only post-menopausal women who have undergone hysterectomy can be considered for such therapy. For the great majority of women who require oestrogen treatment for contraception or as hormone replacement therapy, it must be combined with a progestogen. The final balance of favourable and unfavourable effects on lipid metabolism in any individual will then depend on the preparation used. Secondly oestrogens do increase the risk of thromboembolism and like other steroids will have some mineralocorticoid and glucocorticoid activity, thus increasing the tendency to hypertension and diabetes mellitus.

Many other drugs affect lipoprotein metabolism.[6] Of importance are the retinoic acid derivatives used in dermatology and also, because of the high rate of atherosclerosis in recipients of renal transplants, corticosteroids and cyclosporin. Of great theoretical interest are drugs and chemicals, which induce hepatic microsomal enzymes dependent on cytochrome P450, because of an associated increase in serum HDL levels. Such drugs include phenytoin, phenobarbitone, rifampicin, and griseofulvin. Industrial chemicals with the same effect are chlorinated pesticides such as lindane and DDT.

LIVER DISEASE

Cholestasis

In obstructive jaundice without severe hepatocellular dysfunction there is hypercholesterolaemia. This is due to an increase in unesterified cholesterol in particles of hydrated density similar to LDL. There may also be a moderate hypertriglyceridaemia and an increase in the plasma lecithin concentration.[6,67] The lipoproteins of density similar to LDL are not true apolipoprotein B-containing LDL, levels of which may be low, but are predominantly another lipoprotein designated lipoprotein X (LpX). This contains free cholesterol and phospholipid in an approximately equal molar ratio. LpX has a lamellar structure and on electron microscopy appears as rouleaux of stacked disc-like vesicles. LpX comprises

6% protein, of which half or more is albumin enclosed within the vesicles. Apolipoproteins especially apolipoprotein C are present on their surface. LpX migrates towards the cathode on electrophoresis on agar (but not agarose) which is unusual for a plasma lipoprotein. Its presence in the blood is largely due to the reflux of biliary phospholipids into the circulation, which attract cholesterol out of cell membranes.[67] Biliary cholesterol does contribute to the cholesterol of LpX, but on its own is insufficient to explain the extent of hypercholesterolaemia occurring in many patients. LCAT deficiency also contributes to the accumulation of unesterified cholesterol, but again is unlikely to be the sole cause of the hypercholesterolaemia, because only relatively small quantities of LpX are formed when LCAT activity is even more profoundly decreased in familial LCAT deficiency.

LpX is catabolised by the reticuloendothelial system, including Kupffer cells. Although not itself taken up by the hepatocyte, it may interfere with hepatic uptake of chylomicron remnants.[68] Currently the view is emerging that a system may exist for the sequestration of remnants in the space of Disse before their uptake by the hepatocyte and it is interesting to speculate that this may be a site of their interaction with LpX. It may explain the persistance of remnant-like lipoproteins in patients with obstructive jaundice.

Hepatocellular disease

Moderate hypertriglyceridaemia often accompanies hepatocellular disease. This is due to triglyceride-rich lipoproteins with density in the VLDL and LDL range, but which have β mobility, forming a broad β band on electrophoresis.[6,9,67] The HDL which is present also has β-mobility and when isolated in the ultracentrifuge consists predominantly of small particles. The accumulation of small HDL and the decrease in cholesteryl ester is secondary to LCAT deficiency and the lipoproteins intermediate between VLDL and LDL probably build up because of hepatic lipase deficiency and other damage to the remnant removal mechanism.[6,9,67]

HYPERURICAEMIA AND GOUT

Hyperuricaemia is present in a high proportion (probably half or more), of the men who have hypertriglyceridaemia.[6] Gout is thus frequently met in the Lipid Clinic particularly when hyperuricaemia has been further 'precipitated' by thiazide diuretic administration. The reason for the association is not entirely clear, because it appears to be more common than might be explained by the frequent coincidence of factors, such as obesity and high alcohol consumption, with hypertriglyceridaemia. Hypertriglyceridaemia and hyperuricaemia are not causally related since lowering uric acid with allopurinol does not affect triglyceride levels and conversely with two exceptions lipid lowering drug therapy does not alter the serum urate concentration. The two exceptions are nicotinic acid which raises urate and fenofibrate which decreases it. The latter effect is, however, not mediated through the triglyceride-lowering action of fenofibrate, but because it has an independent uricosuric effect.[69] Both urate and triglyceride levels may decrease on a reducing diet, suggesting that they may both be epiphenomena of some underlying nutritional process. It has been suggested that dietary carbohydrate is important. Dietary fructose which is taken up almost exclusively by the liver induces hypertriglyceridaemia and also increases urate levels,[69] probably by diverting energy away from the hepatic urate scavanging pathway into fructose phosphorylation.

ACKNOWLEDGEMENTS

I am grateful to Mrs A Gharapetian and Mrs J Heydon for expert secretarial assistance.

REFERENCES

1 Durrington PN, Winocour PH. Therapeutic aspects of hyperlipidaemia in diabetes. Postgrad Med J 1989; 65 (Suppl.1): S33–S41
2 Wing AJ, Brunner FP, Brynger H et al. Cardiovascular-related causes of death and the fate of patients with cardiovascular disease. Contrib Nephrol 1984; 41: 306–311
3 Moorhead JF, Chan MK, El-Nahas M, Varghese Z. Lipid nephrotoxicity in chronic progressive glomerular and tubulo-interstitial disease. Lancet 1982; ii: 1309–1311
4 McIntyre N. Plasma lipids and lipoproteins in liver disease. Gut 1978; 19: 526–530
5 Harmony JAK, Akeson AL, McCarthy BM, Morris RE, Scupham DW, Grupp SA. Immunoregulation by plasma lipoproteins. In: Scanu AM, Spector AA, eds. Biochemistry and Biology of Plasma Lipoproteins. New York: Marcel Dekker, 1986: 403–452
6 Durrington PN. Secondary hyperlipidaemia in Hyperlipidaemia Diagnosis and Management. London: Wright, 1989: 219–276
7 Jarrett RJ, Keen H, Chakrabarti R. Diabetes, hyperglycaemia and arterial disease. In: Keen H, Jarrett RJ, eds. Complications of Diabetes. London: Edward Arnold, 1982: 179–204

8 Barclay M. Lipoprotein class distribution in normal and diseased states. In: Nelson GJ ed. Blood Lipids and Lipoproteins. New York: Wiley—Interscience, 1972: 585–704

9 Havel RJ, Goldstein JL. Brown Lipoproteins and lipid transport. In: Bondy PK, Rosenberg LE, eds. Metabolic Control and Disease, 8th Edn. Philadelphia: Saunders, 1980: 393–494

10 Nikkila EA. Triglyceride metabolism in diabetes mellitus. Progr Biochem Pharmacol 1973; 8: 271–299

11 Brunzell JD, Bierman EL. Chylomicronaemia syndrome. Interaction of genetic and acquired hypertriglyceridaemia. Med Clin N Am 1982; 66: 455–468

12 Durrington PN, Newton RS, Weinstein DB, Steinberg D. Effects of insulin and glucose on very-low-density lipoproteins triglyceride secretion by cultured rat hepatocytes. J Clin Invest 1982; 70: 63–73

13 Mangiapane EH, Brindley DN. Effects of dexamethasone and insulin on the synthesis of triacylglycerols and phosphatidyl chlorine and the secretion of very low density lipoproteins and lysophosphatidyl choline by monolayer cultures of rat Hepatocytes. Biochem J 1986; 233: 151–160

14 Pullinger CR, North JD, Teng B-B, Rificia VA, Ronhildde Britto AE, Scott J. The apolipoprotein B gene is constitutively expressed in Hep G2 cells; regulation of secretion by oleic acid, albumin, and insulin, and measurement of the mRNA half-life. J Lipid Res 1989; 30: 1065–1077

15 Field JB. Extraction of insulin by liver. Ann Rev Med 1973; 24: 309–314

16 Greenfield M, Kolferman O, Olefsky J, Reaven GM. Mechanisms of hypertriglyceridaemia in diabetic patients with fasting hyperglycaemia. Diabetologia 1980; 18: 441–446

17 Pietri AO, Dunn FL, Grundy SM, Raskin P. The effect of continuous subcutaneous insulin infusion on very-low-density lipoprotein triglyceride metabolism in Type I diabetes mellitus. Diabetes 1983; 32: 75–81

18 Johnston DG, Alberti KG. Hormone control of ketone body metabolism in the normal and diabetic state. Clin Endocrinol Metabolism 1982; 11: 329–361

19 West KM, Ahuja MM, Bennett PH, et al. The role of circulating glucose and triglyceride concentrations and their interactions with other 'risk factors' as determinants of arterial disease in nine diabetic population samples from the WHO Multinational Study. Diabetes Care 1983; 6: 361–369

20 Winocour PH, Durrington PN, Ishola M, Anderson DC. Lipoprotein abnormalities in insulin-dependent diabetes mellitus. Lancet 1986; i: 1176–1178

21 Steiner G. Hypertriglyceridaemia and carbohydrate intolerance: interrelations and therapeutic implications. Ann J Cardiol 1986; 57: 27G–309

22 Kissebah AM, Alfarsi S, Evans DJ, Adams PW. Integrated regulation of very low density lipoprotein triglyceride and apolipoprotein-B kinetics in non-insulin-dependent diabetes. Diabetes 1982; 31: 217–225

23 Kasama T, Yoshino G, Iwatani I, et al. Increased cholesterol concentration in intermediate density lipoprotein fraction of normolipidaemic non-insulin-dependent diabetes. Atherosclerosis 1987; 63: 263–266

24 Winocour PH, Tetlow L, Durrington PN, Hillier V, Anderson DC. Apolipoprotein E polymorphism and lipoproteins in insulin treated diabetes mellitus. Atherosclerosis 1989; 75: 167–173

25 Curtiss LK, Witztum JL. Plasma apolipoproteins AI, AII, B CI and E are glucosylated in hyperglycaemic diabetic subjects. Diabetes 1985; 34: 452–461

26 Mazzane T, Foster D, Chait A. In vivo stimulation of low-density lipoprotein degradation by insulin. Diabetes 1984; 33: 333–338

27 Chisolm III GM, Morel DW. Lipoprotein oxidation and cytotoxicity: effect of probucol on streptozotocin-treated rats. Am J Cardiol 1988; 62: 20B–26B

28 Brownlee M, Vlassara H, Cerami A. Non-enzymatic glucosylation products on collagen covalently trap low-density lipoprotein. Diabetes 1985; 34: 938–941

29 Durrington PN. Serum high density lipoprotein cholesterol in diabetes mel-

litus: an analysis of factors which influence its concentration. Clin Chim Acta 1980; 104: 11–23

30 Durrington PN, Brownlee WC. Large DM Short-term effects of β-adrenoceptor blocking drugs with and without cardioselectivity and intrinsic sympathomimetic activity on lipoprotein metabolism in hypertriglyceridaemic patients and in normal men. Clin Sci 1985; 69: 713–719

31 Nikkila EA. High density lipoproteins in diabetes. Diabetes 1981; 30 (Suppl 2): 82–87

32 Laakso M, Saarlund H, Ehnholm C, Vouitilainen E, Arno A, Pyorila K. Relationship between postheparin plasma lipases and high density lipoprotein cholesterol in different types of diabetes. Diabetologia 1987; 30: 703–706

33 Winocour PH, Durrington PN, Ishola M, Gordon C, Jeacock J, Anderson DC. Does residual insulin secretion (assessed by C-peptide concentration) affect lipid and lipoprotein levels in insulin-dependent diabetes mellitus? Clin Sci 1989; 77: 369–374

34 Himsworth HP. The dietetic factor determining the glucose tolerance and sensitivity to insulin of healthy man. Clin Sci 1935; 2: 67–94

35 Lalor BC, Bhatnagar D, Winocour PH, et al. Placebo-controlled trial of the effects of guar gum and metformin on fasting blood glucose and serum lipids in obese, type 2 diabetic patients. Diabetic Med 1990; 7: 242–245

36 Winocour PH, Durrington PN, Ishola M, Hillier VF, Anderson DC. The prevalence of hyperlipidaemia and related clinical features in insulin-dependent diabetes mellitus. Qu J Med 1989; 70: 265–276

37 Winocour PH, Durrington PN, Ishola M, Anderson DC, Cohen H. Influence of proteinuria on vascular disease, blood pressure, and lipoproteins in insulin dependent diabetes mellitus. Br Med J 1987; 294: 1648–1651

38 Jones SL, Close CF, Mattock MB, Jarrett RJ, Keen H, Viberti GC. Plasma lipid and coagulation factor concentrations in insulin dependent diabetics with microalbuminuria. Br Med J 1989; 298: 487–490

39 Hart A, Thorp JM, Cohen H. Lipoprotein and fibrinogen studies in diabetes. Postgrad Med J 1971 (June Suppl): 435–439

40 Thompson GR, Soutar AK, Spengel FA, Jadhav A, Gavigan S, Myant NB. Defects of the receptor-mediated low density lipoprotein catabolism in homozygous familial hypercholesterolaemia and hypothyroidism in vivo. Proc Natl Acad Sci USA 1981; 78: 2591–2595

41 Chait A, Kanter R, Green W, Kenny M. Defective thyroid hormone action in fibroblasts cultured from subjects with the syndrome of resistance to thyroid hormones. J Clin Endocrinol Metab 1982; 54: 767–772

42 Abrams JJ, Grundy SM, Gisberg H. Metabolism of plasma triglyceride in hypothyroidism and hyperthyroidism in man. J Lipid Res 1981; 22: 307–322

43 Althaus BU, Staub J-J, Ruff-de Leche A, Oberhansli A, Stahelin HB. LDL/HDL changes in subclinical hypothyroidism: possible risk factors for coronary heart disease. Clin Endocrinol 1988; 28: 157–163

44 Series JJ, Biggart EM, O'Reilly D St J, Packard CJ, Shepherd J. Thyroid dysfunction and hypercholesterolaemia in the general population of Glasgow, Scotland. Clin Chim Acta 1988; 172: 217–222

45 Wieland von H, Seidel D. Plasma lipoprotein bei patienten mit hyperthyreose:isolierung und charakterisierung eines abnormen high-density-lipoproteins. Z Klin Chem Klin Biochem 1972; 10: 311–321

46 Kesaniemi YA, Grundy SM. Increased very low density lipoprotein production associated with obesity. Arteriosclerosis 1983; 3: 170–177

47 Pykalisto OJ, Smith PH, Brunzell JD. Determinants of human adipose tissue lipoprotein lipase: effects of diabetes and diet on basal and diet-induced activity. J Clin Invest 1975; 06: 1108–1117

48 Katan MB. Diet and HDL. In: Miller NE, Miller GJ, eds. Clinical and Meta-

bolic Aspects of High-Density Lipoproteins. Amsterdam: Elsevier 1984: 103–131
49 Chait A, Mancini M, February AW, Lewis B. Clinical and metabolic study of alcoholic hyperlipidaemia. Lancet 1972; ii: 62–63
50 Durrington PN, Twentyman OP, Braganza JM, Miller JP. Hypertriglyceridaemia and abnormalities of triglyceride catabolism persisting after pancreatitis. Int J Pancreatol 1986; 1: 195–203
51 Martin PJ, Martin JV, Goldberg DH. γ-glutamyl transpeptidase, triglycerides and enzyme induction. Br Med J 1975; i: 17–18
52 Chan MK, Varghese Z, Moorhead JF. Lipid abnormalities in uremia, dialysis and transplantation. Kidney Int 1981; 19: 625–637
53 Yedgar S, Weinstein DB, Patsch W, Schonfield G, Casanada FE, Steinberg D. Viscosity of culture medium as a regulator of synthesis and secretion of very low density lipoproteins by cultured hepatocytes. J Biol Chem 1982; 257: 2188–2192
54 Appel GB, Blum CB, Chien S, Kunis Ch, Appel AS. The hyperlipidaemia of nephrotic syndrome. Relation to plasma albumin concentration, oncotic pressure, and viscosity. N Engl J Med 1985; 312: 1544–1548
55 Utermann G, Hoppichler F, Dieplinger H, Seed M, Thompson G, Boerwinkle E. Defects in the 'low density' lipoprotein receptor gene affects lipoprotein (a) levels: multiplicative interaction of two gene loci associated with premature atherosclerosis. Proc Natl Acad Sci USA 1989; 86: 4171–4174
56 Janus ED, Nicoll A, Wooton R, Turner PR, Magill PJ, Lewis B. Quantitative studies of very low density lipoprotein conversion to low density lipoprotein in normal controls and primary hyperlipidaemic states and the role of direct secretion of low density lipoprotein in heterozygous familial hypercholesterolaemia. Eur J Clin Invest 1980; 10: 149–159
57 Short CD, Durrington PN, Mallick NP, Hunt LP, Tetlow L, Ishola M. Serum and urinary high density lipoproteins in glomerular disease with proteinuria. Kidney Int 1986; 29: 1224–1228
58 Nestel PJ, Fidge NH, Tan MH. Increased lipoprotein remnant formation in chronic renal failure. N Engl J Med 1982; 307: 329–333
59 Sniderman A, Cianflone K, Kwiterovich PO, Hutchinson T, Barre P, Pritchard S. Hyperapobetalipoproteinaemia. The major dyslipoproteinaemia in patients with chronic renal failure treated with chronic ambulatory peritoneal dialysis. Atherosclerosis 1987; 65: 257–264
60 Neary RH, Gowland E. The effect of renal failure and haemodialysis on the concentrations of free apolipoprotein A1 in serum and the implications for the catabolism of high density lipoproteins. Clin Chim Acta 1988; 171: 239–246
61 Bhatnagar D, Neary R, Ishola M, Arrol S, Durrington PN. Is prebeta high density lipoprotein the preferred substrate for lecithin:cholesterol acyltransferase? Atherosclerosis 1989; 79: 95
62 Ames RP. The effects of antihypertensive drugs on serum lipids and lipoproteins I Diuretics. Drugs 1986; 32: 260–278
63 Ames RP. The effect of antihypertensive drugs on serum lipids and lipoproteins II Non-diuretic drugs. Drugs 1986; 32: 335–357
64 Miller NE. Effects of adrenoceptor-blocking drugs on plasma lipoprotein concentrations. Am J Cardiol 1987; 60: 17E–23E
65 Durrington PN, Brownlee WC, Large DM. Short-term effects of β-adrenoceptor blocking drugs with and without cardioselectivity and intrinsic sympathomimetic activity on lipoprotein metabolism in hypertriglyceridaemic patients and in normal men. Clin Sci 1985; 69: 713–719
66 Falko JM, Schonfield G, Witztum JL, Kolar J, Weidman SW. Effect of oestrogen therapy on apolipoprotein E in type III hyperlipoproteinaemia. Metabolism 1979; 28: 1171–1177

67 McIntyre N. Plasma lipids and lipoproteins in liver disease. Gut 1978; 19: 526–530
68 Walli AK, Seidel D. Role of lipoprotein-X in the pathogenesis of cholestatic hypercholesterolaemia. Uptake of lipoprotein-X and its effect on 3-hydroxy-3-methylglutaryl coenzyme A reductase and chylomicron remnant removal in human fibroblasts, lymphocytes, and in the rat. J Clin Invest 1984; 74: 867–879
69 Bastow MD, Durrington PN, Ishola M. Hypertriglyceridaemia and hyperuricaemia: effects of two fibric acid derivatives (bezafibrate and fenofibrate) in a double-blind, placebo-controlled trial. Metabolism 1988; 37: 217–220

British Medical Bulletin (1990) Vol. 46, No. 4, pp. 1025–1058
© The British Council 1990

Treatment of hyperlipidaemia

D R Illingworth
Division of Endocrinology, Metabolism and Clinical Nutrition, Department of Medicine, The Oregon Health Sciences University, Portland, Oregon, USA

The primary goal of therapy in the treatment of hyperlipidaemia is to reduce the plasma concentrations of known atherogenic lipoproteins thereby reducing or even reversing the flux of lipids from plasma into the arterial wall. A less common goal is to prevent the adverse sequellae of hyperchylomicronaemia in patients with severe hypertriglyceridaemia. The aetiologic factor(s) responsible for hyperlipidaemia in a given patient need to be clearly established and it is important not to overlook potentially treatable secondary disorders. Diet is the cornerstone of therapy in the treatment of hyperlipidaemia; the decision to begin drug therapy should be individualized and should be made only after an adequate trial of diet has failed to achieve satisfactory concentrations of plasma lipids and lipoproteins. In this review I will discuss the use of diet and drugs in the treatment of patients with hypercholesterolaemia due to increased plasma concentrations of low density lipoproteins, patients with combined hyperlipidaemia in which very low and low density lipoproteins are elevated and patients with severe hypertriglyceridaemia in which chylomicronaemia is present.

WHY SHOULD HYPERLIPIDAEMIA BE TREATED?

Therapeutic interventions directed at modifying plasma concentrations of lipids and lipoproteins in patients identified to have hyperlipidaemia have two goals. One, to reduce the plasma concentrations of known atherogenic lipoproteins, particularly low density lipoproteins (LDL), very low density lipoprotein (VLDL) remnants and Lp(a) and concurrently to increase plasma concen-

0007–1420/90/0046–1025/$10.00

trations of potentially antiatherogenic high density lipoproteins (HDL) therefore exerting a favourable effect upon lipid deposition in the arterial wall. Two, to decrease the plasma concentrations of triglyceride-rich lipoproteins (chylomicrons and VLDL) in patients with severe hypertriglyceridaemia thereby preventing the development of hepatomegaly, splenomegaly, eruptive xanthomas and potentially pancreatitis as well as to reduce the long term risk of atherosclerosis.

The causal relationship between hypercholesterolaemia, particularly that associated with an increase in plasma concentrations of LDL and VLDL remnants and the premature development of atherosclerosis has gained widespread acceptance over the last five years.[1-6] This has been based on evidence from a number of areas including epidemiologic studies, the induction of atherosclerosis in experimental animals, clinical observations of accelerated atherosclerosis in patients with genetic disorders in which plasma concentrations of LDL or VLDL remnants are increased and the reduction in either cardiovascular events or angiographically documented coronary atherosclerosis observed in clinical trials of cholesterol lowering in middle aged men with primary hypercholesterolaemia[7,8] or coronary atherosclerosis.[9] Although areas of controversy still exist, particularly with respect to the treatment of hypercholesterolaemia in older people[10] without evidence of atherosclerosis, in women and in children, the evidence that treatment of hypercholesterolaemia does reduce the risk of atherosclerosis in humans is sufficiently compelling to justify not excluding these patient groups from potential treatment. Evidence for a reduced risk of atherosclerosis from the treatment of hypertriglyceridaemia in both men and women is incomplete.[11,12]

CRITERIA FOR THE DIAGNOSIS OF HYPERLIPIDAEMIA

Epidemiologic studies have demonstrated that lipid and lipoprotein concentrations vary amongst different populations and are generally highest in countries consuming a 'western diet' and lowest in those countries where the habitual consumption of fat and cholesterol is low.[13] There is general agreement[14] that mean plasma cholesterol concentrations in Britain are too high and that a downward shift in the population mean would have a substantial impact on the development of coronary atherosclerosis. Indeed, the majority of cases of coronary heart disease occur in patients with plasma cholesterol concentrations between 5.2 and 7 mM/l and up to a third of cases occur in patients with total plasma cholesterol concentrations below 5.2 mM/l.[15] Determination of the plasma concentrations of cholesterol and triglyceride with a concurrent measurement of HDL cholesterol represents the basic lipid profile necessary for the diagnosis of most hyperlipidaemias. Although a number of dyslipoproteinaemias are now characterized based on apoprotein abnormalities these assays are not in routine use in clinical practice and are available only in specialized centers with a major interest in lipoprotein disorders. Plasma triglyceride concentrations increase postprandially and to avoid spurious increases in triglycerides linked to meals standard procedures call for the patient to have fasted for 12–15 hours prior to venepuncture. In contrast, plasma cholesterol concentrations are minimally affected by eating and casual blood samples are adequate if only cholesterol is to be determined. The importance of determining plasma lipid and lipoprotein concentrations under steady state conditions cannot be overemphasized; lipid concentrations decrease during periods of weight loss and commonly fall for 2–6 weeks following a myocardial infarction.

In evaluating the significance of lipid and lipoprotein concentrations in a given patient it is important to consider the age of the patient and to be familiar with what are the average concentrations for that age and sex as well as the concentrations above which therapeutic interventions with diet and drugs should be considered. Unfortunately, mean concentrations of total and LDL cholesterol among adult subjects in the United Kingdom are higher than desirable and in a range permissive for the development of atherosclerosis. Criteria for the diagnosis of hypercholesterolaemia and, more specifically, for hypercholesterolaemia associated with increased concentrations of LDL cholesterol are therefore somewhat arbitrary and take into account the non-linear relationship between plasma cholesterol concentrations and risk of cardiovascular disease in which the risk increases substantially when plasma cholesterol concentrations exceed 7 mM/l. Mean plasma concentrations of total cholesterol and triglyceride and calculated low density lipoprotein cholesterol in adult men and women in the United Kingdom are shown in Table 1.[14]

In 1985 the Consensus Conference on lowering blood cholesterol to prevent heart disease[1] recommended that severe hypercho-

Table 1   Mean (SD) plasma concentrations (mM/l) of cholesterol, triglyceride and LDL cholesterol by age and sex in Britain

Age group (years)	Total cholesterol Men	Women	Total triglyceride Men	Women	Calculated low density lipoprotein cholesterol Men	Women
25–29	5.2 (1.1)	5.1 (1.0)	1.4 (1.0)	1.1 (0.6)	3.2 (1.1)	3.0 (0.9)
30–34	5.5 (1.1)	5.2 (1.0)	1.6 (1.3)	1.1 (0.6)	3.3 (1.1)	3.1 (1.0)
35–39	5.8 (1.2)	5.3 (1.0)	1.7 (1.2)	1.1 (0.7)	3.6 (1.1)	3.2 (1.0)
40–44	6.0 (1.2)	5.6 (1.1)	1.8 (1.7)	1.2 (1.0)	3.8 (1.1)	3.4 (1.0)
45–49	6.1 (1.2)	5.9 (1.2)	1.9 (1.4)	1.4 (0.9)	3.8 (1.1)	3.7 (1.0)
50–54	6.1 (1.2)	6.4 (1.1)	1.9 (1.5)	1.5 (0.9)	3.9 (1.1)	4.0 (1.1)
55–59	6.1 (1.2)	6.7 (1.2)	1.9 (1.2)	1.7 (1.3)	3.9 (1.1)	4.3 (1.2)

Data from Reference 14

lesterolaemia requiring diet and potentially drug therapy be considered to be present if cholesterol concentrations exceeded the 90th percentile. Moderate hypercholesterolaemia was defined as serum cholesterol concentrations between the 75th and 90th percentiles. Recent expert panels in the United States,[2] Europe,[3] Great Britain[4] and Canada[6] have defined more specific cut points for the diagnosis of hypercholesterolaemia in adults and have suggested levels of LDL cholesterol above which diet and drug therapy should be considered. Under optimal circumstances, treatment decisions should be based upon lipid and lipoprotein concentrations; if only total cholesterol is measured, patients with high levels of HDL cholesterol (greater than 2–2.5 mM/l) will not be identified and may be inappropriately treated for hypercholesterolaemia. The expert committees[2–4,6] have concluded that desirable levels of total cholesterol for adults in western societies are under 5.2 mM/l and that in most patients in this category, further characterization of lipoproteins is not recommended. With the caveat that these recommendations will fail to detect patients with atypically low concentrations of HDL cholesterol whose total cholesterol levels are below 5.2 mM/l, the latter value is an appropriate goal of therapy.

The expert committees[2–4,6] have advocated dietary therapy for patients with total cholesterol concentrations between 5.2 and 6.5 mM/l with more stringent dietary recommendations being advocated for the majority of patients whose total cholesterol values range between 6.5 and 7.8 mM/l. Some patients in the latter category may be candidates for drug therapy. Patients whose total cholesterol concentrations exceed 7.8 mM/l and who have LDL cholesterol concentrations above 4.9 mM/l are at high risk for premature coronary heart disease and such patients require individual therapy with diet and, potentially, drugs. The NCEP Panel[2] has advocated a more aggressive approach to lipid lowering therapy in patients with evidence of atherosclerosis or those who have two other cardiovascular risk factors; in such patients drug therapy should be considered when the LDL cholesterol concentration exceeds 4.2 mM/l on maximal dietary therapy. In patients without evidence of atherosclerosis, the NCEP panel advocated a therapeutic goal of under 4.2 mM/l for LDL cholesterol whereas for individuals with atherosclerosis or the concurrent presence of two risk factors a lower level of LDL (less than 3.4 mM/l) is desirable. Concentrations of LDL cholesterol below 2.5 mM/l may be necessary for the regression of atherosclerosis to occur in humans.[9]

The 1984 National Institutes of Health Consensus Conference on hypertriglyceridaemia[12] pointed out that there is a continuous distribution of triglyceride concentrations amongst patients in western industrialized societies which tend to increase with age and parallel the rise in adiposity. Treatment of hypertriglyceridaemia remains controversial[3,11,12] and the selection of a particular triglyceride concentration above which intervention is necessary remains arbitrary. Mild to moderate hypertriglyceridaemia (2.5–6 mM/l) in the absence of concurrent hypercholesterolaemia is commonly attributable to secondary disorders which should be vigorously addressed. In the author's opinion, drug therapy for patients with this degree of hypertriglyceridaemia is rarely indicated and should be tailored to the individual patient; it is most appropriate for patients with genotypic familial combined hyperlipidaemia in whom the risk of premature coronary artery disease is increased. Hypertriglyceridaemia is also seen in patients with combined elevations of VLDL and LDL (phenotypic Type IIB hyperlipidaemia) or in patients with Type III hyperlipidaemia in whom VLDL remnant particles accumulate in plasma. Aggressive diet and potentially drug therapy is appropriate for patients with these disorders when plasma cholesterol concentrations exceed 6.5–8 mM/l with concurrent triglyceride levels of between 3–8 mM/l. Vigorous dietary and, in selected cases, drug therapy is appropriate for patients with triglyceride concentrations exceeding 10–12 mM/l in whom the goal of therapy is to decrease triglycerides to under 3–5 mM/l with the elimination of chylomicronaemia and its associated risk of abdominal pain and pancreatitis.

EVALUATION OF THE PATIENT WITH HYPERLIPIDAEMIA

On the basis of initial determinations of the concentrations of total plasma (or serum) cholesterol and triglyceride hyperlipidaemias can be divided into four categories: (1) hypercholesterolaemia with normal concentrations of triglyceride; (2) combined elevations of cholesterol and triglycerides in which plasma triglyceride levels are one to three times higher than the cholesterol concentration (Type IIB and III phenotypes); (3) conditions associated with a primary elevation in the concentration of triglyceride-rich lipoproteins (VLDL) and in which cholesterol concentrations are normal or are very mildly increased (Type IV phenotype); and (4) conditions in which triglyceride concentrations are markedly elevated with a concurrent increase in plasma cholesterol and in which the plasma looks lipaemic (Type I and Type V phenotypes). In adult patients, severe hypertriglyceridaemia is invariably associated with increased plasma concentrations of chylomicrons and VLDL particles. Determination of the plasma concentrations of HDL cholesterol will identify those patients with hypercholesterolaemia in whom this is attributable to high levels of HDL (>2–2.5 mM/l) and in the absence of concurrent increases in LDL cholesterol these patients should be reassured and require no specific therapy. A fifth category of patient with dyslipidaemia attributable to atypically low plasma concentrations of HDL cholesterol will not be identified on the basis of elevated concentrations of total cholesterol or triglycerides[16,17] but appears to be at increased risk of coronary artery disease.[15,18,19] Potential approaches to the treatment of these patients will be discussed briefly later in this review.

Treatment of hyperlipidaemia generally implies a commitment to long term therapy and it is important that the goals of such therapy be clear to both the treating physician and the patient. The importance of establishing the causal factor(s) responsible for hyperlipidaemia in a given patient cannot be overemphasized; the characterization of, and distinction between, primary and secondary causes of hyperlipidaemia has been discussed elsewhere in this issue (Thompson, Durrington). Although certain secondary causes of hyperlipidaemia respond to correction of the primary underlying disorder (e.g. treatment of primary hypothyroidism), in other disorders the primary abnormality may not be amenable to treatment and the secondary hyperlipidaemia must be managed by the use of diet and, in selected patients, by the concurrent use of specific lipid lowering drugs. Concentrations of plasma lipids and lipoproteins may also be influenced by a number of prescription drugs, as discussed in the Chapter on Secondary Hyperlipidaemia, and it is important to consider these as potential exacerbating causes of hyperlipidaemia in a given patient.

DIETARY TREATMENT OF HYPERLIPIDAEMIA

Dietary modification represents the initial therapy for all patients with hyperlipidaemia including those who concurrently require drug therapy. The main dietary changes are aimed at reducing plasma concentrations of LDL cholesterol and, in patients with concurrent moderate hypertriglyceridaemia to decrease plasma triglycerides and the concentrations of VLDL cholesterol. In general, dietary changes exert their lipid lowering effects by influencing either the hepatic synthesis of VLDL or the expression of high affinity LDL receptors in the liver. The latter are decreased by dietary saturated fatty acids (possibly excluding stearic acid) and cholesterol.[20,21] The main dietary modifications needed to reduce plasma LDL cholesterol concentrations involve decreased intakes of dietary cholesterol and total fat (with a particular emphasis on the saturated fat content) and, in patients with hypertriglyceridaemia or those who are overweight, the avoidance of excess calories particularly those derived from alcohol. Prior to dietary modification, a typical patient may be consuming 40–45% of total calories from fat of which 15–20% are from saturated fat with a daily cholesterol intake of 450–600 mg. Dietary modification aims to reduce the total fat intake to less than 30% of calories, saturated fat to less than 10% of calories and the daily intake of cholesterol to below 300 mg/day.[2,3] Further reductions in the saturated fat content to less than 7% of total calories and in the daily intake of cholesterol to below 200 mg/day (the step 2 diet) will produce a further lowering of LDL cholesterol in most patients. Replacement of saturated fats in the diet is best achieved by increasing the intake of complex carbohydrates but moderate increases in the content of monounsaturated fatty acids and omega-6 polyunsaturated fatty acids may also be beneficial.[22–24] Additional dietary intakes of fish oils enriched in omega-3 fatty acids have a triglyceride-lowering effect (at daily intakes exceeding 5 g of omega-3 fatty acids per day) but, in patients with hypercholesterolaemia due to increased concentrations of LDL cholesterol or combined hyperli-

Table 2   Dietary therapy of hypercholesterolaemia

Nutrient	Step-one diet	Step-two diet
Total Fat*	Less than 30%	Less than 30%
Fatty acids*		
Saturated	Less than 10%	Less than 7%
Polyunsaturated	Up to 10%	Up to 10%
Monounsaturated	10% to 15%	10% to 15%
Carbohydrates*	50% to 60%	50% to 60%
Protein*	10% to 20%	10% to 20%
Cholesterol	Less than 300 mg/day	Less than 200 mg/day
Total calories	To achieve and maintain desirable weight	To achieve and maintain desirable weight

(From Ref. 2) *Percent of total calories

pidaemia, such supplements may raise LDL cholesterol concentrations[25,26] and cannot be recommended.

The dietary modifications outlined in Table 2 are practical and primarily involve substitution rather than complete changes in the eating style of the individual patient. The magnitude of lipid reduction achieved by these dietary changes depends on the initial dietary habits of the patient, his or her initial levels of total and LDL cholesterol and individual diet responsiveness. If the patient has been consuming a typical British diet containing 40–45% of calories from fat with a cholesterol intake of 500 mg/day, dietary modification to the step one diet may result in plasma cholesterol reductions of 0.5 to 1 mM/l.[27] Further dietary modifications to the Step 2 diet may result in an additional 0.2–0.4 mM/l decrease.

It is important to allow an adequate trial of dietary therapy (typically 3–6 months) to enable the patient and his family to make the suggested changes in his or her lifestyle. Consultation with a dietitian is particularly helpful and other family members should be involved in the nutritional education and counselling sessions. Their support is invariably helpful to the patient and they in turn are likely to benefit from increased knowledge of a diet that has been recommended for the general population.

In addition to the hyperlipidaemic effects of excess caloric intake, saturated fats and cholesterol, a number of other dietary components have been reported to influence plasma concentrations of lipids and lipoproteins.[28] Thus, mild hypercholesterolaemia has been associated with the drinking of boiled, but not filtered, coffee[29] whereas a modest hypocholesterolaemic effect has been observed in some[30] but not all[31] studies in which soluble fibers have been incorporated into a cholesterol lowering diet. The frequency of meals may also influence plasma lipid concentrations; a hypocholesterolaemic effect has been observed in response to nibbling as compared to gorging.[32]

The dietary treatment of severe hypertriglyceridaemia involves, acutely the elimination of exogenous fat with the concurrent reduction in total calories. Once chylomicronaemia has been reduced, progressive increases in the fat content up to 15–25% of total calories can be introduced. Restriction of alcohol and, in overweight patients, caloric restriction coupled with an increase in exercise may be particularly helpful in the long term management of these patients. Medium chain triglycerides, which are not incorporated into chylomicron particles, may be useful and, as will be discussed later, fish oils rich in omega-3 fatty acids may also have a role in the dietary therapy of selected patients whose triglycerides remain above 10–12 mM/l.[33]

DRUG TREATMENT OF PRIMARY HYPERCHOLESTEROLAEMIA

The decision to institute drug therapy for hyperlipidaemia implies a long-term commitment to such therapy; this is most likely to succeed if the rationale and goals of therapy are clear to both the prescribing physician and the patient. Factors to be considered in the decision to use drugs to reduce LDL cholesterol concentrations include the magnitude, duration, and aetiology of hypercholesterolaemia, the family history of premature coronary artery disease, the age and sex of the patient, the presence or absence of atherosclerosis, the concurrent presence of other risk factors for coronary artery disease (e.g. low levels of HDL, diabetes), the attitude of the patient towards drug therapy and the potential benefit to be derived from such treatment. Factors to be considered in choosing a drug for initial therapy in adult patients with primary hypercholesterolaemia include the magnitude of hypercholesterolaemia, the age and lifestyle of the patient, relative contraindications to first choice drugs, and potentially, the overall costs of therapy. In the opinion of the author, the bile acid resins, cholestyramine and colestipol, nicotinic acid, simvastatin, and bezafibrate are the drugs of first choice for initial therapy of most adult patients with primary hypercholesterolaemia with the first three being the most effective (Table 3). Bile acid sequestrants and nicotinic acid were advocated as drugs of first choice by the Adult Treatment Panel[2] and both of these drugs have been shown to

Table 3 Drug therapy of primary hypercholesterolaemia

Lipoprotein abnormality	Plasma lipoprotein elevated	Drug therapy	
		First choice agents	Second-line drugs
Primary Hypercholesterolaemia (FH, FCH, undefined)	LDL	Cholestyramine Colestipol Nicotinic acid Lovastatin Simvastatin (Pravastatin) Bezafibrate (Fenofibrate) (Ciprofibrate) Combined drug therapy	Probucol Gemfibrozil Acipimox

Abbreviations: LDL = low density lipoproteins, VLDL = very low density lipoproteins, FH = familial hypercholesterolaemia, FCH = familial combined hyperlipidaemia, (not approved)

reduce cardiovascular morbidity and mortality and have a long established record of clinical use. In the author's opinion, however, simvastatin also warrants inclusion as a drug of first choice for treatment of adult patients with primary hypercholesterolaemia (excluding women of child bearing potential) because of its potent ability to lower LDL cholesterol and relatively low incidence of side effects in short and moderately long term use.[34,35]

Cholestyramine and colestipol have been extensively evaluated in well conducted clinical trials and, in compliant patients, these drugs lower plasma concentrations of LDL cholesterol by 15–30%.[34] The dose-response curves for both of these drugs are, however, non-linear and in adult patients, a 10–20% decrease in LDL cholesterol can often be achieved with doses of 10 g/day of colestipol or 8 g/day of cholestyramine. It is important to counsel patients on the need to take these drugs with meals and initially to start with a low dose (1–2 scoops/day). Maximal doses of colestipol (30 g/day) or cholestyramine (24 g/day) are rarely justified and the modest further decrease (4–8%) in LDL cholesterol which can be obtained when the dose of cholestyramine is increased from 16 g/day to 24 g/day is more than offset by a reduction in patient compliance and an increase in gastrointestinal side effects. The primary action of both cholestyramine and colestipol is to bind bile acids in the intestinal lumen with a concurrent interruption in the enterohepatic circulation of bile acids and a markedly increased excretion of faecal steroids. Depletion of the bile acid pool stimulates an increase in the hepatic synthesis of bile acids from cholesterol, and decreases the hepatic pool of cholesterol with a resultant compensatory increase in cholesterol biosynthesis and in the number of high affinity LDL receptors expressed on hepatocyte membranes. The increased number of high affinity LDL receptors stimulates an increased rate of LDL catabolism and results in a decrease in the concentrations of these lipoproteins in plasma.[36] The compensatory increase in hepatic cholesterol biosynthesis which occurs during bile acid sequestrant therapy limits the overall hypolipidaemic effect of this class of drugs and is frequently paralleled by an increase in hepatic triglyceride and VLDL production with a resultant increase in plasma triglyceride concentrations. The magnitude of this increase may be greater in patients with familial combined hyperlipidaemia than in patients with heterozygous familial hypercholesterolaemia.[37] Bile acid sequestrants have the potential to reduce LDL cholesterol concentrations to less than 3.5 mM/l in patients whose initial values are in the 4.5–5.5 mM/l range and such changes are commonly paralleled by a 3–8% increase in the plasma concentrations of HDL cholesterol.

Common side effects observed with the bile acid sequestrants include changes in bowel function and the potential to exacerbate pre-existing haemorrhoids. Cholestyramine and colestipol are not absorbed from the gastro-intestinal tract and do not require detailed monitoring for potential biochemical side effects. Cholestyramine and colestipol do, however, have the potential to bind certain other co-administered drugs such as digitalis, thyroxine, thiazide diuretics, and warfarin and may decrease the absorption of folic acid and fat soluble vitamins. Bile acid sequestrants do not reduce plasma concentrations of Lp(a).[38]

Nicotinic acid (niacin) is an effective and inexpensive drug for the treatment of patients with primary hypercholesterolaemia and is probably the drug of initial choice for those patients with familial combined hyperlipidaemia.[2,37] The hypocholesterolaemic effects of nicotinic acid are unrelated to its action as a coenzyme in intermediary metabolism and are mediated by a decreased hepatic synthesis of VLDL and LDL.[39,40] When used as a hypolipidaemic agent, nicotinic acid has been shown to reduce cardiovascular morbidity and mortality and total mortality.[41] Nicotinic acid is available in regular and sustained-release preparations; the latter are associated with a lower incidence of flushing but, on a gram for gram basis, are inferior in their ability to lower LDL cholesterol concentrations[42] and are more hepatotoxic.[43] Patients should be advised not to change from regular to sustained-release formulations of nicotinic acid due to the increased risk of hepatotoxicity at the same daily dose.[43] Niacinamide, which is widely available in health food stores, has no effect on plasma lipid or lipoprotein concentrations.

Therapy with nicotinic acid is usually initiated with a dose of 100–250 mg daily with food and the dose then progressively increased up to a dose of 1.5 to 2 g/day in divided doses. Further dose increases up to a maximum of 4.5–6 g/day may be made in incremental amounts but, after each dosage increase, lipid values together with tests of liver function and uric acid should be determined. Patients should be advised to take nicotinic acid with or after meals to minimize the cutaneous flushing but the latter side effects become less troublesome with prolonged therapy. The flushing is prostaglandin mediated and can be inhibited by concurrent administration of small doses of aspirin or nonsteroidal anti-inflammatory drugs. When used as a single agent at doses of 3–6 g/day, nicotinic acid reduces plasma concentrations of LDL cholesterol by 20–35%, raises the levels of HDL by 10–20% and reduces plasma triglycerides by 20–40%.[37,42] Rises in the plasma concentrations of HDL cholesterol are seen in response to low doses of nicotinic acid (0.75–1.5 g/day) but higher doses 2–6 g/day) are necessary to reduce LDL cholesterol concentrations.[37,44] In addition to its ability to reduce LDL cholesterol concentrations, nicotinic acid at a dose of 4 g/day has also been reported to reduce Lp(a) concentrations by 38%.[45]

Side effects occur commonly in patients treated with nicotinic acid and the drug cannot be tolerated by all patients. In addition to the predictable cutaneous flushing, side effects which often detract from effective long-term therapy include: nausea, abdominal discomfort, dryness of the skin, and rarely, blurred vision.[42–46] Nicotinic acid should be regarded as contraindicated in patients with active liver disease, a prior history of peptic ulcer disease, or hyperuricaemia and is relatively contraindicated in patients with Type II diabetes. Laboratory abnormalities are more common in patients treated with doses of >3 g/day of nicotinic acid and include increases in the serum concentrations of uric acid, glucose, aminotransferases and alkaline phosphatase. These biochemical parameters need to be monitored periodically during therapy with nicotinic acid. Although successful use of nicotinic acid requires considerable skill on the part of the prescribing physician and motivation on the part of the patient, this drug is an extremely effective and inexpensive medication and favorably influences the plasma concentrations of all lipoproteins.

Acipimox, a nicotinic acid analogue, at a dose of 250 mg three times daily has been shown to reduce LDL cholesterol concentrations by 11% with a concurrent increase in HDL of 20% in patients with primary hypercholesterolaemia.[47] Acipimox therefore appears to be less effective than nicotinic acid and cannot be recommended as a first choice drug for the treatment of primary hypercholesterolaemia.

Simvastatin is the first of a new class of drugs to be approved in Britain which act as specific inhibitors of 3 hydroxy-3 methyl glutaryl co-enzyme A (HMG CoA) reductase, the rate limiting enzyme in cholesterol biosynthesis.[34,48] Several other drugs in this class have either been approved by regulatory authorities in other countries (e.g. lovastatin) or are undergoing clinical trials (e.g. pravastatin, fluvastatin). Lovastatin is made by a soil fungus (*aspergillus terrius*) and simvastatin is a methylated derivative of lovastatin. The structure of these drugs is shown in Figure 1. Lovastatin and simvastatin are administered as lactones and hydrolysis to the open acid form occurs in the liver; in contrast, pravastatin is taken as the open acid. Data on the relative rates of uptake of these drugs by the liver and peripheral tissues are contradictory[49,\50] and further work is needed to assess whether the greater hydrophylicity of pravastatin as compared with simvastatin and lovastatin results in lower rates of uptake of this drug by peripheral tissues of hypercholesterolaemic patients. At clinically effective doses, however, all these drugs are primarily taken up by the liver and

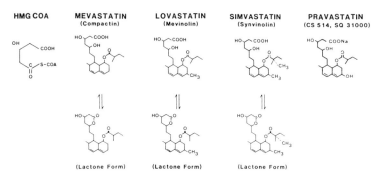

Fig. 1 Structural similarities between HMG CoA and some of the recently developed HMG CoA reductase inhibitors.

exert their hypolipidaemic effects by inhibiting hepatic HMG CoA reductase.

The hypocholesterolaemic effects of lovastatin and simvastatin have been most thoroughly examined and, in patients with heterozygous familial hypercholesterolaemia, both drugs reduce the rate of formation of mevalonic acid by 30–35% at doses of 40 mg/day.[51,52] Reduced formation of mevalonic acid leads to a corresponding decrease in hepatic cholesterol biosynthesis with a reduction in the cellular pool of cholesterol. This leads to a compensatory increase in the number of high affinity LDL receptors expressed on the cell surface which, in turn, stimulates an increase in receptor mediated catabolism of VLDL remnants and LDL.[53] Hepatic synthesis of VLDL and LDL may concurrently be reduced in response to drug therapy with HMG CoA reductase inhibitors.[54]

As a class, HMG CoA reductase inhibitors are the most potent of the available drugs for reducing LDL cholesterol concentrations.[55] On a mg for mg basis, simvastatin appears to be more potent than either lovastatin or pravastatin and decreases of 28%, 30–38%, and 37–44% have been observed in the plasma concentrations of LDL cholesterol in patients with heterozygous familial hypercholesterolaemia treated with doses of 10, 20, and 40 mg/day of simvastatin in single or divided doses.[34,52,56,57] Simvastatin is equally effective given as a single dose in the evening as compared to a twice daily regimen but, because cholesterol biosynthesis in humans increases at night, the drug is more effective when given as a single dose in the evening as compared to a similar dose given in the morning. Therapy with simvastatin and lovastatin has been associated with a 20–30% decrease in the plasma concentrations of triglycerides and an overall tendency for HDL cholesterol to increase from 2–15%.[34,48,52,56,57] Despite considerable individual variability in hypolipidaemic response, which is independent of apoprotein E phenotype,[58] lovastatin or simvastatin appear to be equally effective in reducing LDL cholesterol concentrations in patients with heterozygous familial hypercholesterolaemia as compared to patients with familial combined hyperlipidaemia or other less well characterized disorders.[35,59,60] Despite their potent ability to reduce LDL cholesterol concentrations, neither lovastatin nor simvastatin reduce plasma concentrations of Lp(a)[61] and, in some patients, the concentrations of Lp(a) have been reported to increase slightly.[61]

Simvastatin at doses of 10–40 mg/day and lovastatin at doses of

20–80 mg/day have usually been well tolerated and side effects in short and moderately long term (up to 7 years) have been fairly uncommon.[35,59–60,62] Reported side effects to date have included changes in bowel function, headaches, nausea, fatigue, insomnia, skin rashes, and myopathy. Development of myopathy does not appear to be due to an unusual susceptibility to inhibition of mevalonic acid production by lovastatin[63] and the aetiologic factor(s) responsible for this remain poorly defined. Myopathy appears to be uncommon in patients in lovastatin or simvastatin as mono therapy (incidence < 1 in 500); in contrast the incidence of myopathy is increased in patients concurrently receiving cyclosporine, nicotinic acid, gemfibrozil (and potentially other fibrates), and erythromycin[64] and HMG CoA reductase inhibitors should be used very cautiously in combination with these other drugs. Simvastatin, lovastatin and their metabolites are excreted primarily in bile and toxic levels could easily be reached if these drugs were given to patients with cholestasis or other disorders in which there is a potential impairment in hepatic excretion or metabolism. Initial pre-clinical studies in dogs[64] indicated the potential of HMG CoA reductase inhibitors to cause cataracts when administered in doses of 20–80 times the maximal human dose; despite initial concern about this potential side effect in humans, close monitoring of patients maintained on lovastatin or simvastatin for periods of 3–5 years has failed to document any increased frequency of lenticular opacities[35,64] and initial concerns that these drugs would cause cataracts in humans have not been substantiated. Biochemical tests to monitor the safety of simvastatin should be assessed at 6–8 week intervals for the first 9–12 months of therapy and at 3–4 monthly intervals thereafter. These should primarily include assessments of liver and muscle function and possibly electrolytes. Serum glucose concentrations are not affected by simvastatin or lovastatin.

A number of drugs structurally related to clofibrate have been developed which are more potent in lowering LDL cholesterol concentrations than is clofibrate. Drugs in this class include bezafibrate, fenofibrate, and ciprofibrate. Although not strictly a clofibrate analogue, gemfibrozil is also commonly discussed in the same class as the other fibrates. On the basis of its ability to lower LDL cholesterol concentrations, bezafibrate, at a dose of 600 mg/day, justifies inclusion as one of the more effective LDL lowering drugs in the treatment of primary hypercholesterolaemia. Fenofibrate and ciprofibrate are of similar efficacy and warrant inclusion as potential first choice drugs in those countries where they are available, whereas gemfibrozil and clofibrate are less effective in reducing LDL cholesterol concentrations and cannot be recommended as first line therapy. Reductions in the plasma concentrations of LDL cholesterol of 18–28%[65,66] have been observed in patients with primary hypercholesterolaemia treated with bezafibrate (600 mg/day); similar reductions have been reported with ciprofibrate (100 mg/day) or fenofibrate (300 mg/day).[34,37] In contrast, decreases of only 4.6 and 9.6% respectively have been observed in the concentrations of LDL cholesterol in patients with heterozygous familial hypercholesterolaemia treated with either clofibrate (2 gm/day) or gemfibrozil (1200 mg/day).[34,37] This indicates that clofibrate and gemfibrozil should not be considered as initial agents for single drug therapy in patients with heterozygous familial hypercholesterolaemia. When used in patients with primary hypercholesterolaemia, bezafibrate therapy results in a 10–20% increase in the plasma concentrations of HDL cholesterol and a 20–40% decrease in plasma triglycerides.[65–67]

The hypolipidaemic effects of the fibric acid class of drugs result from a number of different mechanisms including reductions in the synthesis of VLDL triglycerides and, to a lesser extent, VLDL apoprotein B, increased activity of lipoprotein lipase and an enhanced rate of receptor mediated clearance of LDL from plasma.[68,69]

Bezafibrate is generally well tolerated; the most common minor side effects include: changes in bowel function, abdominal pain, headaches, and (rarely) a rash. Rhabdomyolysis has been reported when the drug has been used in patients with renal insufficiency.[70] Biochemical changes in patients treated with bezafibrate have included increases in aminotransferases, and creatine kinase, and a decrease in alkaline phosphatase. Bezafibrate, like other fibrate drugs, has a potential to increase biliary lithogenicity with a predicted long term increase in the incidence of gallstones.[71]

In addition to bile acid sequestrants, nicotinic acid, simvastatin, and bezafibrate which constitute the most effective hypolipidaemic drugs for the treatment of primary hypercholesterolaemia, several other lipid lowering medications are available that may be of use in selected patients (Table 3). Probucol is a modestly effective drug for treating primary hypercholesterolaemia and at the recommended dose of 500 mg twice daily, reduces LDL cholesterol concentrations by 8–15%.[72] Individual patient response to probucol is quite variable, but the drug appears to be equally effective in patients with heterozygous familial hypercholesterolaemia as compared to those with other less well characterized causes of primary hypercholesterolaemia.[73,74] The precise mechanism(s) by which probucol lowers LDL cholesterol concentrations is unclear but the drug appears to increase non-receptor mediated catabolism of LDL.[75] Probucol has no effect on plasma triglycerides but reduces the concentrations of HDL by up to 25%.[72] Despite its modest LDL lowering effect, however, probucol has been shown to cause regression of tendon xanthomas in patients with both homozygous and heterozygous familial hypercholesterolaemia[76] and may have other potentially beneficial effects on lipoproteins including inhibition of LDL oxidation[77] and an increase in reverse cholesterol transfer.[78] Probucol has not been utilized in any primary or secondary prevention trials published to date and it is unclear whether or not the antioxidant effects of this drug may contribute to a reduction in atherosclerosis in patients with hypercholesterolaemia. For this reason, the precise role of probucol in the management of hyperlipidaemia is controversial; based on its modest ability to lower LDL concentrations, however, I consider it as a second line drug in the treatment of primary hypercholesterolaemia.

Probucol is well tolerated and side effects occur in less than 5% of patients. These include: changes in bowel function, abdominal pain, and nausea. Probucol is stored in adipose tissue and blood levels decrease slowly following its discontinuation. It would seem prudent not to use the drug in female patients who plan future pregnancies. Probucol causes prolongation of the QT interval and should probably be regarded as contra-indicated in patients with initially long QT intervals or a history of ventricular arrhythmias.[72]

Two other drugs, neomycin and d-thyroxine, have been used in the treatment of primary hypercholesterolaemia but, in my opinion, cannot be recommended. Neomycin at a dose of 1 g twice daily has been shown to reduce LDL cholesterol concentrations by 10–15% in patients with primary hypercholesterolaemia.[79] Changes in bowel function including diarrhoea and abdominal cramps occur in 30–40% of patients treated with this drug and it has the potential to cause impairment in renal function or ototoxicity if given to patients with inflammatory bowel disease or a potential for increased gastro-intestinal absorption. D-thyroxine has been shown to reduce concentrations of LDL cholesterol by

10–15% but does so at the expense of making the patient modestly hyperthyroid.[80] Use of this drug is not recommended.

COMBINATION DRUG THERAPY FOR PRIMARY HYPERCHOLESTEROLAEMIA

The vast majority of patients with primary hypercholesterolaemia who are treated with drugs will be satisfactorily controlled on monotherapy with one of the agents of first choice. For patients with severe hypercholesterolaemia, particularly those with heterozygous familial hypercholesterolaemia, the use of two or more drugs with different mechanisms of actions has been effectively utilized in several combined drug regimens.[35,37,81] The efficacy of some of the more established combinations are summarized in Table 4. The bile acid sequestrants (cholestyramine and colestipol) enhance fecal steroid excretion and are not systemically acting; they have provided the cornerstone for the majority of the established combined drug regimens. The most effective of these have used a bile acid sequestrant in combination with nicotinic acid, simvastatin, or lovastatin. Combinations in which bezafibrate or probucol have been used with cholestyramine or colestipol have also been shown to be useful, although the LDL lowering effect of probucol appears to be quite variable. The use of low doses of bile acid sequestrants (8 g/day of cholestyramine or 10 g/day of colestipol) in combination with lovastatin or simvastatin provides a therapeutic regimen that is usually well tolerated, shows additive lipid lowering, and is cost effective.[81] Reductions in LDL cholesterol concentrations of 60–70% have been observed in patients

Table 4   Efficacy of combined drug regimens in severe primary hypercholesterolaemia

Drug combination	Percent decrease in LDL cholesterol	Number of patients studied	Reference
Cholestyramine + Nicotinic Acid	−48	6	85
Colestipol + Nicotinic Acid	−47	11	86
Cholestyramine + Bezafibrate	−31	18	65
Colestipol + Lovastatin	−54	10	87
Cholestyramine + Simvastatin	−54	13	60
Cholestyramine + Simvastatin	−64	5	88
Lovastatin + Nicotinic Acid	−49	8	34
Lovastatin + Colestipol + Nicotinic Acid	−67	21	82

with severe heterozygous familial hypercholesterolaemia treated with the combination of lovastatin, colestipol, and nicotinic acid.[34,82] Certain drug combinations are not recommended due to either a lack of efficacy or potential for increased toxicity; these include combined drug therapy with acipimox and cholestyramine,[83] or the combination of lovastatin (or simvastatin) and probucol,[74] or gemfibrozil.[84] Although the latter combination may have potential therapeutic appeal in patients with combined hyperlipidaemia, it does not significantly reduce LDL cholesterol concentrations (as compared to lovastatin alone) in patients with familial hypercholesterolaemia and is associated with an increased risk of myopathy.[84] Referral to a consultant or centre specializing in the management of lipid disorders is strongly recommended for the management of most patients with severe hypercholesterolaemia in whom combination drug therapy is needed.

NON-PHARMACOLOGICAL TREATMENT OF SEVERE HYPERCHOLESTEROLAEMIA

Several surgical techniques, plasmaphaeresis and more selective forms of aphaeresis which utilize extracorporeal adsorbents have been used in the treatment of patients with both homozygous and heterozygous familial hypercholesterolaemia; their use, however, is limited and should be pursued only after detailed consultation with a specialist in lipoprotein disorders. Distal ileal bypass surgery reduces LDL cholesterol concentrations by 30–40% in patients with heterozygous familial hypercholesterolaemia[89] but is ineffective in homozygous patients. With the availability of potent LDL lowering drugs such as simvastatin, this operation cannot be recommended. Liver transplantation has been effectively used in the therapy of children with homozygous familial hypercholesterolaemia[90] and affords a physiologically effective means of providing functional LDL receptors to these patients. This operation appears to be more effective than porto caval shunt surgery.[91] Repetitive plasma exchange performed at 2 week interval has been successfully used to treat patients with homozygous familial hypercholesterolaemia and results in improved survival of this patient group.[92] Technical advances in aphaeresis have led to the development of selective extracorporeal adsorbents which remove only LDL without concurrently removing other plasma proteins or HDL. Several adsorbents have been used in these columns including heparin, dextran sulphate, or antibodies against apoprotein

B;[93-95] the efficiency of these procedures can be improved by using 2 columns and an automatic regenerating system which avoids saturation of the column.[96] Concentrations of LDL cholesterol can be reduced to under 1 mM/l by these techniques but rise rapidly in the ensuing days thereby necessitating repeat aphaeresis in 7–10 days in patients with homozygous familial hypercholesterolaemia or 10–15 days in severe heterozygous patients. LDL aphaeresis is the treatment of choice for patients with homozygous familial hypercholesterolaemia in whom this technique is fully justified; its use in other patient populations should be considered investigational.

DRUG THERAPY OF COMBINED HYPERLIPIDAEMIA

Increased plasma concentrations of cholesterol and triglycerides can be due to the presence of an increased number of LDL and VLDL particles of normal composition (phenotypic Type IIB hyperlipidaemia) or may reflect the presence of abnormal cholesterol-enriched VLDL particles which accumulate in the plasma of patients with Type III hyperlipidaemia. The goals of drug therapy for the former patients are to reduce the concentrations of LDL cholesterol with a secondary aim being to reduce plasma triglyceride, the concentrations of VLDL cholesterol and potentially increase HDL cholesterol levels. In patients with Type III hyperlipidaemia, the goal is to reduce the concentrations of atherogenic VLDL remnant particles.

Correction of potentially exacerbating secondary factors is important in patients with combined hyperlipidaemia and dietary or environmental factors that enhance VLDL production should be sought and, if feasible, modified. Nicotinic acid is a drug of first choice in patients with combined hyperlipidaemia (Table 5) and in doses of 3–6 g/day reduces VLDL concentrations by 30–80% with a concurrent reduction of 10–30% in the plasma concentrations of LDL cholesterol.[37] HDL cholesterol levels frequently rise by 10–25%. Reductions of 25–45% in the plasma concentrations of LDL cholesterol have been observed in patients with combined hyperlipidaemia treated with lovastatin at doses of 20 or 40 mg twice daily[35,97] or with simvastatin at doses of 20 or 40 mg daily.[59] These changes were accompanied by 5–10% increases in HDL cholesterol and a 20–30% decrease in plasma triglycerides.[35,59,97,98] The ability of HMG CoA reductase inhibitors to favorably affect concentrations of VLDL, LDL, and HDL

Table 5 Drug therapy of combined hyperlipoproteinaemia

Lipoprotein abnormality	Plasma lipoproteins elevated	Drug therapy	
		First choice agents	Second-line drugs
Combined Hyperlipoproteinaemia (FCH, FH, undefined)	LDL+ VLDL	Nicotinic acid Lovastatin Simvastatin (Pravastatin) Bezafibrate Gemfibrozil (Fenofibrate) Combined drug therapy	Cholestyramine Colestipol Clofibrate Acipimox
Type III Hyperlipoproteinaemia	Chylomicron and VLDL remnants	Clofibrate Bezafibrate Gemfibrozil (Fenofibrate) Lovastatin Simvastatin Nicotinic acid	

Abbreviations: LDL = low density lipoproteins, VLDL = very low density lipoproteins, FH = familial hypercholesterolaemia, FCH = familial combined hyperlipidaemia

makes these drugs attractive agents to use in the therapy of patients with combined hyperlipidaemia who are unable to take nicotinic acid. Bezafibrate and gemfibrozil are effective in reducing plasma concentrations of triglycerides and VLDL cholesterol in patients with combined hyperlipidaemia but their effects on LDL concentrations are most unpredictable.[66,99] Bezafibrate appears to be more effective in reducing LDL cholesterol concentrations than is gemfibrozil and decreases of 15–30% may be observed with the former drug[66]; in some patients, however, plasma concentrations of LDL cholesterol may increase during therapy with either bezafibrate or gemfibrozil.[66,99] Both fibrates increase plasma concentrations of HDL by 15–25%; in the case of gemfibrozil, treatment of patients with combined hyperlipidaemia was associated with a reduction in cardiovascular morbidity and mortality in the Helsinki Heart Trial.[8] Cholestyramine and colestipol increase concentrations of plasma triglycerides and VLDL cholesterol when administered to patients with combined hyperlipidaemia and should be avoided as initial therapeutic agents in these patients. Insufficient data is available to recommend the use of either probucol or acipimox in patients with combined hyperlipidaemia although the latter would, on theoretical grounds, be expected to have some potential therapeutic usefulness.

Combination drug therapy to treat patients with combined hyperlipidaemia has received less attention than for the treatment of
severe hypercholesterolaemia but appears to have both theoretical
and practical advantages. Combined drug therapy with nicotinic
acid and a bile acid sequestrant or with either bezafibrate or gemfibrozil and a bile acid sequestrant will both reduce concentrations
of VLDL and LDL in plasma and raise the levels of HDL cholesterol.[99] Combination therapy with lovastatin or simvastatin and a
hypotriglyceridaemic agent such as nicotinic acid, gemfibrozil, or
bezafibrate may also be predicted to be an effective drug combination for the treatment of patients with combined hyperlipidaemia,[99] however these combinations appear to be associated with
the potential for an increased risk of myopathy[64] and cannot be
generally recommended.

Type III hyperlipidaemia is an uncommon disorder characterized by the presence of abnormal cholesterol-rich VLDL and,
in some patients, chylomicron remnant particles in plasma.
Patients with this disorder respond well to dietary changes, weight
loss, and correction of potentially exacerbating secondary factors
but in patients whose cholesterol concentrations remain above
6.5–7 mM/l, drug therapy is appropriate. Reductions of 40–60%
in the plasma concentrations of cholesterol and triglyceride can be
obtained in patients with Type III hyperlipidaemia in response to
treatment with bezafibrate, gemfibrozil, or clofibrate.[100,101] Nicotinic acid,[102] lovastatin,[103] and simvastatin[62] are also effective in
the therapy of Type III hyperlipidaemia whereas the condition is
exacerbated if bile acid sequestrants are prescribed. Oestrogen
therapy may also be useful in the management of postmenopausal
women with Type III hyperlipidaemia.[104]

DRUG THERAPY OF HYPERTRIGLYCERIDAEMIA

The preceding section has discussed treatment of patients with
combined hyperlipidaemia and here I will focus on management
of patients with endogenous hypertriglyceridaemia in which
plasma concentrations of VLDL particles are increased (Type IV
phenotype) and patients with more severe hypertriglyceridaemia
in which both chylomicron particles and VLDL are present in
excess (phenotypic Type V hyperlipidaemia).

Moderate hypertriglyceridaemia (3–7 mM/l) in the absence of
concurrent hypercholesterolaemia is commonly seen in patients
with coronary atherosclerosis and, in one study of 500 survivors of

myocardial infarction, was present in 15%.[105] Most patients with endogenous hypertriglyceridaemia will concurrently have reduced levels of HDL cholesterol and it has been difficult to delineate the potential benefits from therapy which reduces plasma triglyceride concentrations but concurrently increases the concentrations of HDL cholesterol.[106] In the opinion of the author, drug therapy for patients with endogenous hypertriglyceridaemia should be considered only after correction of potentially exacerbating secondary factors (body weight, alcohol) has failed to reduce triglycerides to below 4–5 mM/l and, in such patients, is most appropriate for those with familial combined hyperlipidaemia, a strong family history of premature coronary artery disease or in those patients who have themselves developed clinical evidence of coronary or peripheral vascular disease at a relatively young age. Nicotinic acid is the drug of choice for these patients and at doses of 2–4.5 g/day reduces VLDL cholesterol concentrations by 50–75% with a concurrent increase in HDL cholesterol levels of 15–30%. Concentrations of LDL cholesterol in these patients are initially low and may increase slightly in response to nicotinic acid therapy. Gemfibrozil or bezafibrate are reasonable second choice drugs for patients with endogenous hypertriglyceridaemia and, like nicotinic acid, reduce triglycerides and plasma concentrations of VLDL cholesterol by 50–80% with a concurrent 10–30% increase in HDL cholesterol. However, the magnitude of increase in LDL cholesterol concentrations during fibrate therapy is greater than that seen with nicotinic acid and may result in LDL cholesterol concentrations exceeding 3–4 mM/l.[66,99] The use of fish oil preparations (e.g. MaxEPA®) is not recommended in the treatment of patients with endogenous hypertriglyceridaemia and invariably results in increased plasma concentrations of LDL cholesterol.[107,108]

Drug therapy is appropriate for patients with severe hypertriglyceridaemia in whom plasma triglyceride concentrations remain above 10–11 mM/l after appropriate correction of secondary factors and in conjunction with dietary therapy (Table 6). Fish oil preparations rich in omega-3 fatty acids (e.g. MaxEPA®) are also effective in the treatment of severe hypertriglyceridaemia but when used in these patients, their use must be regarded as a pharmaceutical agent rather than a dietary supplement. Doses of fish oil preparation providing 5 g/day of omega-3 fatty acids have been reported to reduce plasma triglycerides by 50–60% in patients with severe hypertriglyceridaemia;[109] supplements providing 20–30 g/day are even more effective and reduce triglyceride con-

Table 6   Drug therapy of hypertriglyceridaemia

Lipoprotein abnormality	Plasma lipoproteins elevated	Drug therapy	
		First choice agents	Second-line drugs
Severe Hypertriglyceridaemia	Chylomicrons and VLDL	Gemfibrozil Nicotinic acid (omega-3 fatty acids)	Bezafibrate Clofibrate

VLDL = very low density lipoproteins

centrations by 75–80%.[33] The hypotriglyceridaemic effect of these long chain polyunsaturated fatty acids is mediated by a reduction in VLDL production and by a decrease in postprandial lipaemia.[25,110] Although fish oil preparations such as MaxEPA® have a potential role in the treatment of selected patients with severe hypertriglyceridaemia, their use may be associated with a deterioration in diabetic control due to an increase in the hepatic synthesis of glucose[111] and they frequently contribute an unwanted source of additional calories.

In my opinion, gemfibrozil at a dose of 600 mg twice daily is the drug of first choice for patients with severe hypertriglyceridaemia and use of this agent lowers plasma triglyceride concentations by 70–90%, often with a complete elimination of chylomicronaemia.[112] These changes are paralleled by increases in the plasma concentrations of HDL cholesterol and by modest increases in LDL cholesterol.[112] Bezafibrate and clofibrate are also effective hypotriglyceridaemic drugs but appear less effective in the treatment of patients with severe hypertriglyceridaemia than is gemfibrozil.[67,112,113] Nicotinic acid represents a good second choice drug for the treatment of patients with chylomicronaemia who do not have diabetes but the efficacy of this drug appears to be less predictable than gemfibrozil.[114,115] HMG CoA reductase inhibitors do not have therapeutic benefit in patients with chylomicronaemia nor does probucol. The anabolic steroid, oxandrolone, has occasionally been used in male patients with severe hypertriglyceridaemia, as has norethindrone acetate in female patients, but neither can be generally recommended.

TREATMENT OF THE PATIENT WITH A LOW HDL CHOLESTEROL

The strong inverse relationship between plasma concentrations of HDL cholesterol (or apoprotein AI) and the incidence of coronary artery disease observed in western populations is well established.[15,17,18] Concentrations of HDL cholesterol are however also low in societies consuming a high carbohydrate low fat diet in whom the incidence of coronary heart disease is low. Notably in the latter populations, plasma concentrations of LDL cholesterol are generally below 2.5 mM/l and suggests that the inverse correlations noted between low concentrations of HDL cholesterol and high rates of coronary heart disease in western societies may require the concurrent presence of permissively high level of LDL cholesterol (e.g. > 3 mM/l). Low levels of HDL cholesterol occur on a primary basis and are commonly seen in association with secondary factors including hypertriglyceridaemia, obesity, cigarette smoking, physical inactivity and a number of drugs. Despite the strong epidemiologic evidence between low levels of HDL cholesterol and an increased incidence of coronary heart disease it is unclear whether or not therapeutic measures to increase HDL cholesterol levels in individuals in whom these are low are of therapeutic benefit.[116] The use of gemfibrozil in the Helsinki Heart Trial[8] was associated with a reduction in triglycerides and a concurrent increase in HDL which, over a 5 year period, was associated with a significant reduction in cardiovascular morbidity and mortality. This study supports the use of lipid lowering drugs in selected patients with hyperlipidaemia and concurrent low levels of HDL but, in the opinion of the author, the major focus from such therapy should be to reduce plasma concentrations of known atherogenic lipoproteins with drugs which concurrently may raise HDL concentrations. Nicotinic acid and the fibrate drugs gemfibrozil and bezafibrate are the preferred drugs for this use. At the present time, however, the use of pharmaceutical agents which aim **solely** to increase HDL cholesterol levels cannot be justified but, in patients with low levels of HDL, non-pharmaceutical approaches (e.g. weight loss, increase in exercise, cessation of cigarette smoking) should be encouraged. Although low doses of nicotinic acid (0.5–1 g/day) increases HDL cholesterol levels by 0.1–0.4 mM/l in patients with hypoalphalipoproteinaemia,[44] neither gemfibrozil nor lovastatin exert a significant HDL raising effect in patients with isolated hypoalphalipoproteinaemia who do not concurrently have hypertriglyceridaemia.[117] Although proof of benefit remains to be established it is reasonable to attempt to reduce levels of LDL cholesterol to under 2.5 mM/l in patients with persistently low concentrations of HDL cholesterol, particularly in those with premature coronary heart disease or a strongly positive family history.

DRUG THERAPY OF PATIENTS WITH SECONDARY HYPERLIPIDAEMIA

Hyperlipidaemia is commonly seen as a secondary manifestation of endocrine, renal or hepatic diseases and in some of these disorders the primary disorder is not amenable to correction. Premature atherosclerosis is commonly seen in patients with the nephrotic syndrome or renal insufficiency and is the major cause of death in patients with both Type I and Type II diabetes. Recent reports have documented the efficacy of lovastatin and simvastatin in the treatment of patients with secondary hyperlipidaemia associated with the nephrotic syndrome[118-120] and use of these drugs appears justified in patients with persistent nephrotic syndrome and LDL cholesterol levels above 5–6 mM/l. Recent studies have indicated that lovastatin[121] and pravastatin[122] are both effective in reducing concentrations of atherogenic lipoproteins in patients with Type II diabetes and these drugs do not adversely affect diabetic control. This is in contrast to nicotinic acid which tends to exacerbate hyperglycaemia. Bezafibrate and gemfibrozil may also be useful in the treatment of diabetic patients with hyperlipidaemia particularly that associated with increased plasma triglyceride concentrations. Proof of benefit from the treatment of hyperlipidaemia in diabetic patients or those with the nephrotic syndrome has not been obtained but, in view of the causal association between hyperlipidaemia and accelerated rates of atherosclerosis which occur in these patients their hyperlipidaemia should not be ignored. This is particularly true in the case of diabetics.[123]

THERAPY OF HYPERLIPIDAEMIA IN CHILDREN

The increased awareness of hypercholesterolaemia as a risk factor for premature atherosclerosis in adults together with the emerging consensus that detection and treatment of hyperlipidaemia may have greatest benefit if started in childhood[124] has lead to an increased detection of lipid disorders in children.[125] Mean concentrations of total and LDL cholesterol in children are 4 and 2.5 mM/l respectively but cut points for the diagnosis of hyperlipidaemia remain arbitrary and are based on percentiles (e.g. for children aged 5–15 the 90th percentile for LDL cholesterol is 3.5 mM/l). As with adults, in any child identified to have hyperlipidaemia it is important to first delineate the aetiology of the hyperlipidaemia with treatment being directed as correction of secondary factors and, in most children, dietary therapy. Treatment of children identified to have heterozygous familial hypercholesterolaemia involves primarily a low fat low cholesterol diet. Guidelines for the use of lipid lowering drugs in children are incomplete[126] but, in the opinion of the author, should be limited to cholestyramine or colestipol which are not systemically acting. The use of lipid lowering drugs before the age of 5 or 6 is not recommended and the decision on when to begin drug therapy is based on several factors including the magnitude of hypercholesterolaemia, the family history of premature atherosclerosis, the sex of the child (treat boys more than girls) and the attitude of the child and his or her parents. Drug therapy must be individualized but is most appropriate for those children with LDL cholesterol concentrations which exceed 5–6 mM/l on maximal dietary therapy. Supplements of folic acid and fat soluble vitamins are advisable in children maintained on long term bile acid sequestrant therapy. Insufficient data is available concerning the efficacy and particularly the long term safety of simvastatin, lovastatin or bezafibrate to recommend their use in children with heterozygous familial hypercholesterolaemia and the use of any systemically acting drugs in this population should be regarded as investigational. Referral to a centre specializing in lipid disorders is mandatory for the effective management of the rare child with homozygous familial hypercholesterolaemia.[90–92]

Dietary therapy is the mainstay of treatment for children with hypertriglyceridaemia, including those with type I hyperlipidaemia due to lipoprotein lipase deficiency.[126] Drug therapy is rarely indicated for the treatment of children with either hypertriglyceridaemia or combined hyperlipidaemia.

ACKNOWLEDGEMENTS

This work was supported in part by National Institutes of Health Research Grants HL28399, HL37940 and by the General Clinical Research Center's Program (RR334).

REFERENCES

1 Consensus Conference: Lowering blood cholesterol to prevent heart disease. J Am Med Assoc 1985; 253: 2080–2086
2 The Expert Panel. Report of the National Cholesterol Education Program Expert Panel on detection, evaluation and treatment of high blood cholesterol in adults. Arch Intern Med 1988; 148: 36–69
3 Study Group of the European Atherosclerosis Society. The recognition and

management of hyperlipidaemia in adults. A policy statement of the European Atherosclerosis Society. Eur Heart J 1988; 9: 571–600

4   Shepherd J, Betteridge DJ, Durrington P, et al. Strategies for reducing coronary heart disease and desirable limits for blood lipid concentrations: guidelines from the British Hyperlipidaemia Association. Br Med J 1987; 295: 1245–1246

5   The British Cardiac Society Working Group on Coronary Prevention: Conclusions and Recommendations. Br Heart J 1987; 57: 188–189

6   Basinski A, Frank JW, Naylor CD, Rachlis MM. Detection and management of asymptomatic hypercholesterolaemia. A policy document by the Toronto Working Group on cholesterol policy. Toronto: Ontario Ministry of Health, 1989

7   Lipid Research Clinics Program. The Lipid Research Clinics coronary primary prevention trial results 1 and 2. J Am Med Assoc 1984; 251: 351–374

8   Frick MH, Elo O, Haapa K, et al. Helsinki Heart Study: Primary Prevention Trial with gemfibrozil in middle-aged men with dyslipidemia. N Engl J Med 1987; 317: 1237–1245

9   Blankenhorn DH, Nessum SA, Johnson RL, San Marco ME, Azen SP, Cashin-Hemphill L. Beneficial effects of combined colestipol and niacin therapy on coronary atherosclerosis and coronary venous bypass grafts. J Am Med Assoc 1987; 257: 3233–3240

10 Benfante R, Reed D. Is elevated serum cholesterol level a risk factor for coronary heart disease in the elderly? J Am Med Assoc 1990; 263: 393–396

11 Hulley SB, Rosenman RH, Bawal RD, et al. Epidemiology as a guide to clinical decisions. The association between triglyceride and coronary heart disease. N Engl J Med 1980; 302: 1383–1389

12 National Institutes of Health Consensus Conference: Treatment of Hypertriglyceridemia. J Am Med Assoc 1984; 251: 1196–1200

13 Stamler J. Population Studies. In: Levy RI, Rifkind BM, Dennis BH, Ernst N (eds). Nutrition, lipids and coronary heart disease: a global view. New York: Raven Press, 1979: 25–88

14 Mann JI, Lewis B, Shepherd J, et al. Blood lipid concentrations and other cardiovascular risk factors: distribution, prevalence and detection in Britain. Br Med J 1988; 296: 1702–1706

15 Miller M, Mead LA, Kwiterovich PO, Pierson TA. Dyslipidaemias with desirable plasma total cholesterol levels and angiographically demonstrated coronary artery disease. Am J Cardiol 1990; 64: 1–5

16 Grundy SM, Goodman DS, Rifkind BM, Cleeman JI. The place of HDL in cholesterol management. A perspective from the National Cholesterol Education Program. Arch Int Med 1989; 149: 505–510

17 Olson RE. A critique of the report of the National Institutes of Health Expert Panel on detection, evaluation and treatment of high blood cholesterol. Arch Intern Med 1989; 149: 1501–1503

18 Abbott RD, Wilson BWF, Kannel WB, Castelli WP. High density lipoprotein cholesterol, total cholesterol screening and myocardial infarction. Arteriosclerosis 1988; 8: 207–211

19 Livshitz J, Weisbort J, Meshulam N, Brunner D. Multivariate analysis of the 20 year follow up of the Donolo-Tel Aviv Prospective Coronary Artery Disease Study and the usefulness of high density lipoprotein cholesterol percentage. Am J Cardiol 1989; 63: 676–681

20 Spady DK, Dietschy JM. Dietary saturated triacyl glycerols suppress hepatic low density lipoprotein receptor activity in the hamster. Proc Natl Acad Sci USA 1985; 82: 4526–4530

21 Spady DK, Dietschy JM. Interaction of dietary cholesterol and triglycerides in the regulation of hepatic low density lipoprotein transport in the hamster. J Clin Invest 1988; 81: 300–309

22 Becker N, Illingworth DR, Alaupovic P, Sundberg EE, Connor WE. Effects of saturated, monounsaturated, and omega-6 polyunsaturated fatty acids on plasma lipids, lipoproteins and apoproteins in humans. Am J Clin Nutr 1983; 37: 355–360

23 Grundy SM. Comparison of monounsaturated fatty acids and carbohydrates for lowering plasma cholesterol. N Engl J Med 1986; 314: 745–748

24 Mattson FH, Grundy SM. Comparison of effects of dietary saturated, monounsaturated and polyunsaturated fatty acids on plasma lipids and lipoproteins in men. J Lipid Res 1985; 26: 194–202

25 Harris WS. Fish oils and plasma lipid and lipoprotein metabolism in humans. A critical review. J Lipid Res 1989; 30: 785–807

26 Wilt TJ, Lofgren RP, Nichol KL, et al. Fish oil supplementation does not lower plasma cholesterol in men with hypercholesterolemia. Results of a randomized placebo controlled crossover study. Ann Intern Med 1989; 111: 900-905

27 Hatcher LF, Flavell DP, Illingworth DR. Dietary therapy of hypercholesterolemia. Practical cardiology (special issue) May 1988, 31–37

28 Connor WE, Connor SL. Diet, atherosclerosis and fish oil. Adv Intern Med 1990; 35: 139–172

29 Bak AAA, Grobee D. The effect on serum cholesterol levels of coffee brewed by filtering or boiling. N Engl J Med 1989; 321: 1432–1437

30 Kay RM, Truswell AS. Effect of citrus pectin on blood lipids and fecal steroid excretion in men. Am J Clin Nutr 1977; 30: 171–175

31 Swain JF, Rouse IL, Curley CB, Sacks FM. Comparison of the effects of oat bran and low fiber wheat on serum lipoprotein levels and blood pressure. N Engl J Med 1990; 322: 147–152

32 Jenkins DJA, Wolever TMS, Vuksan V, et al. Nibbling vs. gorging: metabolic advantages of increased meal frequency. N Engl J Med 1989; 321: 929–934

33 Phillipson BE, Rothrock DW, Connor WE, Harris WS, Illingworth DR. Reduction of plasma lipids, lipoproteins and apoproteins by dietary fish oil in patients with hypertriglyceridaemia. N Engl J Med 1985; 312: 1210–1216

34 Illingworth DR, Bacon S. Treatment of heterozygous familial hypercholesterolaemia with lipid lowering drugs. Arteriosclerosis 1989; 9: 129–134

35 Illingworth DR, Bacon SP, Larson KK. Long term experience with HMG CoA reductase inhibitors in the therapy of hypercholesterolaemia. Atherosclerosis Rev 1988; 18: 161–187

36 Shepherd J, Packard CJ, Bicker S, et al. Cholestyramine promotes receptor mediated low density lipoprotein catabolism. N Engl J Med 1980; 302: 1219–1222

37 Illingworth DR. Drug therapy of hypercholesterolaemia. Clin Chem 1988; 33: B123–B132

38 Vessby B, Costner G, Lithell H, Thomis J. Diverging effects of cholestyramine on apoprotein B and lipoprotein Lp(a). Atherosclerosis 1982; 44: 61–71

39 Levy RI, Langer T. Hypolipidemic drugs and lipoprotein metabolism. Adv Exp Med Biol 1972; 27: 155–163

40 Grundy SM, Mock HYI, Zech LE, Berman M. The influence of nicotinic acid on metabolism of cholesterol and triglycerides in man. J Lipid Res 1981; 22: 24–36

41 Canner PL, Berge KG, Wanger NK, et al. Fifteen year mortality in coronary drug project patients: long term benefit with niacin. J Am Coll Cardiol 1986; 8: 1245–1255

42 Knopp RH, Ginsberg J, Albers JJ, et al. Contrasting effects of unmodified and time release forms of niacin on lipoproteins in hyperlipidaemic subjects: Clues to mechanism of action of niacin. Metabolism 1985; 34: 642–650

43 Mullin GE, Greenson JK, Mitchell MC. Fulminant hepatic failure after ingestion of sustained release nicotinic acid. Ann Intern Med 1989; 111: 253–255

44   Alderman JD, Pasternak RC, Sacks FM, Smith HS, Monrad ES, Grossman W. Effect of a modified, well tolerated niacin regimen on serum total cholesterol, high density lipoprotein cholesterol, and the cholesterol:high density ratio. Am J Cardiol 1989; 64: 725–729

45   Carlson LA, Hamsten A, Asplund A. Pronounced lowering of serum levels of lipoprotein Lp(a) in hyperlipidaemic subjects treated with nicotinic acid. J Intern Med 1989; 226: 271–276

46   Millay RH, Klein ML, Illingworth DR. Niacin maculopathy. Ophthalmology 1988; 95: 930–936

47   Sirtori CS, Gianfranceschini G, Sirtori M, et al. Reduced triglyceridaemia and increased high density lipoprotein cholesterol levels after treatment with acipimox, a new inhibitor of lipolysis. Atherosclerosis 1981; 38: 267–271

48   Grundy SM. HMG CoA reductase inhibitors for treatment of hypercholesterolaemia. N Engl J Med 1988; 319: 24–33

49   Tsujita Y, Kuroda M, Shimada Y. CS 514, a competitive inhibitor of 3-hydroxy-3 methyl glutaryl coenzyme A reductase: tissue selective inhibition of sterol synthesis and hypolipidaemic effect in various animal species. Biochim Biophys Acta 1986; 877: 50–60

50   Alberts AW. Discovery, biochemistry and biology of lovastatin. Am J Cardiol 1988; 62: 10J–15J

51   Pappu AS, Illingworth DR, Bacon S. Reduction in plasma low density lipoprotein cholesterol and urinary mevalonic acid by lovastatin in patients with heterozygous familial hypercholesterolaemia. Metabolism 1989; 38: 542–549

52   Hagamenas FC, Pappu AS, Illingworth DR. The effects of simvastatin on plasma lipoproteins and cholesterol homeostatis in patients with heterozygous familial hypercholesterolaemia. Eur J Clin Invest 1990; 20: 150–157

53   Brown MS, Goldstein JL. Receptor mediated pathway for cholesterol homeostasis. Science 1986; 232: 34–47

54   Ginsberg HN, Le NA, Short NP, Ramakrishnan R, Desnick RJ. Suppression of apolipoprotein B production during treatment of cholesterol ester storage disease with lovastatin: Implications for regulation of apolipoprotein B synthesis. J Clin Invest 1987; 80: 1692–1697

55   Maher VMG, Thompson GR. HMG CoA reductase inhibitors as lipid-lowering agents: five years experience with lovastatin and an appraisal of simvastatin and pravastatin. Q J Med 1990; 74 (New Series): 165–175

56   Weisweiler P, Schwandt P. Colestipol plus fenofibrate versus sinvinolin in familial hypercholesterolaemia. Lancet 1986; 2: 1212–1213

57   Mol MJTM, Erkelens DW, Gevers-Leuven JA, Schouten JA, Stalenhoef FH. Simvastatin (MK733): A potent cholesterol synthesis inhibitor in heterozygous familial hypercholesterolaemia. Atherosclerosis 1988; 69: 131–137

58   O'Malley JP, Illingworth DR. The influence of apolipoprotein E phenotype on the response to lovastatin therapy in patients with heterozygous familial hypercholesterolaemia. Metabolism 1990; 39: 150–154

59   Ytre-Arne K, Nordoy A. Simvastatin and cholestyramine in the long term treatment of hypercholesterolaemia. J Intern Med 1989; 226: 285–290

60   Lintott CJ, Scot RS, Nye ER, Robertson NC, Sutherland WHF. Simvastatin: An effective treatment for hypercholesterolaemia. Aust NZ J Med 1989; 19: 317–320

61   Kostner GM, Gabish D, Leopold D, Bolzano K, Weintraub MS, Breslow JL. HMG CoA reductase inhibitors lower LDL cholesterol without reducing Lp(a) levels. Circulation 1989; 80: 1313–1319

62   Stalenhoef AFH, Mol MJTM, Stuyt PMJ. Efficacy and tolerability of simvastatin. Am J Med 1989; 87: 39S–43S

63   Maher VMG, Pappu AS, Illingworth DR, Thompson GR. Plasma mevalonate response in lovastatin related myopathy. Lancet 1989; 2: 1098

64 Tobert JA. Efficacy and long term adverse effect pattern of lovastatin. Am J Cardiol 1988; 62: 28J–34J
65 Curtis LD, Dickson AC, Ling KLE, Betteridge J. Combination treatment with cholestyramine and bezafibrate for heterozygous familial hypercholesterolaemia. Br Med J 1988; 297: 173–175
66 Eisenberg S, Gavish D, Kleinman Y. Bezafibrate. In: Fears R (ed). Pharmacological control of hyperlipidemia, Barcelona. JR Prouse Science Publishers, 1986: pp. 145–169
67 Monk JP, Todd PA. Bezafibrate: A review. Drugs 1987; 33: 539–576
68 Kesaniemi YA, Grundy SM. Influence of gemfibrozil and clofibrate on metabolism of cholesterol and plasma triglyceride in man. J Am Med Assoc 1984; 251: 2241–2247
69 Stewart JM, Packard CJ, Lorimer AR, Boag DE, Shepherd J. Effects of bezafibrate on receptor mediated and receptor independent low density lipoprotein catabolism in type II hyperlipoproteinaemic subjects. Atherosclerosis 1982; 44: 355–368
70 Heidemann H, Bock KD, Kruzfelder E. Rhabdomyolysis and bezafibrate therapy in renal insufficiency. Klin Wochenschr 1981; 59: 413–414
71 Leiss O, Bergmann K, Gannso A, Augustine J. Effects of gemfibrozil on biliary lipid metabolism in normolipidaemic subjects. Metabolism 1985; 34: 74–82
72 Buckley MMT, Goa KL, Price AH, Brogden RN. Probucol: A reappraisal of its pharmacological properties and therapeutic use in hypercholesterolaemia. Drugs 1989; 37: 761–800
73 Helve E, Tikkanen MJ. Comparison of lovastatin and probucol in treatment of familial and nonfamilial hypercholesterolaemia: Different effects on lipoprotein profiles. Atherosclerosis 1988; 72: 189–197
74 Witztum JL, Simmons D, Steinberg D, et al. Intensive combination drug therapy of familial hypercholesterolaemia with lovastatin, probucol, and colestipol hydrochloride. Circulation 1989; 79: 16–28
75 Nestel PJ, Billington T. Effects of probucol on low density lipoprotein removal and high density lipoprotein synthesis. Atherosclerosis 1981; 38: 203–209
76 Yamamoto A, Matsuzawa Y, Yokoyama S, et al. Effects of probucol on xanthoma regression in familial hypercholesterolaemia. Am J Cardiol 1986; 57: 29H–35H
77 Steinberg D, Parthasarathy S, Carew TE, Khoo JC, Witztum JL. Beyond cholesterol: Modifications of low density lipoprotein that increase its atherogenicity. N Engl J Med 1989; 320: 915–924
78 Franceschini T, Sirtori M, Vaccarino V, et al. Mechanisms of HDL reduction after probucol: Changes in HDL subfractions and increased reverse cholesterol ester transfer. Arteriosclerosis 1989; 9: 462–469
79 Illingworth DR. Lipid lowering drugs: An overview of indications and optimum therapeutic use. Drugs 1987; 33: 259–279
80 Bantle JP, Oppenheimer JH, Schwartz HL, et al. TSH response to TRH in euthyroid hypercholesterolaemic patients treated with graded doses of dextrothyroxine. Metabolism 1981; 30: 63–68
81 Illingworth DR. New horizons in combination drug therapy for hypercholesterolaemia. Cardiology 1989; 76: 83–100
82 Malloy MJ, Kane JP, Kunitake ST, et al. Complimentarity of colestipol, niacin, and lovastatin in treatment of severe familial hypercholesterolaemia. Ann Intern Med 1987; 107: 616–623
83 Gylling H, VanHanen H, Miettinen TA. Effects of acipimox and cholestyramine on serum lipoproteins, non-cholesterol sterols and cholesterol absorption and elimination. Eur J Clin Pharmacol 1989; 37: 111–115
84 Illingworth DR, Bacon S. Influence of lovastatin plus gemfibrozil on plasma

lipids and lipoproteins in patients with heterozygous familial hypercholestero-laemia. Circulation 1989; 79: 590–596

85 Packard CJ, Steward JM, Morgan HG, Lorimer AR, Shepherd J. Combined drug therapy for familial hypercholesterolaemia. Artery 1980; 7: 281–289

86 Illingworth DR, Phillipson BE, Rapp JH, Connor WE. Colestipol plus nic-otinic acid in treatment of heterozygous familial hypercholesterolaemia. Lan-cet 1981; 1: 296–298

87 Illingworth DR. Mevinolin plus colestipol in therapy for severe heterozygous familial hypercholesterolaemia. Ann Intern Med 1984; 101: 598–604

88 Mölgaard J, VonSchenck H, Olsson AG. Comparative effects of simvastatin and cholestyramine in treatment of patients with hypercholesterolaemia. Eur J Clin Pharmacol 1989; 36: 445–460

89 Spengle FA, Jadhav A, Duffield RGM, et al. Superiority of partial ileal bypass over cholestyramine in reducing cholesterol in familial hypercholesterolaemia. Lancet 1981; 2: 768–771

90 Bilheimer DW, Goldstein JL, Grundy SM, Starzl TE, Brown MS. Liver transplantation to provide low density lipoprotein receptors and lower plasma cholesterol in a child with homozygous familial hypercholesterolaemia. N Engl J Med 1984; 311: 1658–1664

91 Starzl TE, Chase HP, Ahrens EH, Jr., et al. Porta caval shunt in patients with familial hypercholesterolaemia. Ann Surg 1984; 198: 273–282

92 Thompson GR, Miller JP, Breslow JL. Improved survival of patients with homozygous familial hypercholesterolaemia treated with plasma exchange. Br Med J 1985; 291: 1671–1673

93 Lupien PJ, Moorjani S, Awad J. A new approach to the management of familial hypercholesterolaemia: Removal of plasma cholesterol based on the principle of affinity chromatography. Lancet 1976; 1: 1261–1265

94 Yokoyama S, Hayashi R, Satami M, et al. Selective removal of low density lipoprotein by plasmapheresis in familial hypercholesterolaemia. Arterioscler-osis 1985; 5: 613–622

95 Stoffel W, Demant T. Selective removal of apolipoprotein B containing serum lipoproteins from blood plasma. Proc Natl Acad Sci USA, 1981; 78: 611–615

96 Mabuchi H, Michishita I, Takeda M, et al. A new low density lipoprotein apheresis system using two dextran sulphate cellulose columns in an auto-mated column regenerating unit (LDL continuous apheresis). Atherosclerosis 1987; 68: 19–25

97 Tikkanen MJ, Helve E, Jaattela A and the Finish Lovastatin Study Group. Comparison between lovastatin and gemfibrozil in the treatment of primary hypercholesterolaemia: The Finnish Multicenter Study. Am J Cardiol 1988; 62: 35J–43J

98 Tikkanen MJ, Bokanagra TS, Walker JF, Cook T. Comparison of low doses of simvastatin and gemfibrozil in the treatment of elevated plasma cholesterol: A multicenter study. Am J Med 1989; 87 (Suppl 4A): 47S–53S

99 East C, Bilheimer DW, Grundy SM. Combination drug therapy for familial combined hyperlipidaemia. Ann Intern Med 1988; 109: 25–32

100 Stuyt PJM, Stallenhoef AFH, Demacker PMN, et al. A comparative study of the effects of acipimox and clofibrate in type III and type IV hyperlipoprotei-naemia. Atherosclerosis 1985; 55: 51–62

101 Houlston R, Quiney J, Watts GF, Lewis B. Gemfibrozil in the treatment of resistant familial hypercholesterolaemia and type III hyperlipoproteinaemia. J R Soc Med 1988; 81: 274–276

102 Hoogwerf BJ, Bantle JP, Kuba K, et al. Treatment of type III hyperlipoprotei-naemia with four different treatment regimens. Atherosclerosis 1984; 51: 251–259

103 Illingworth DR, O'Malley JP. The hypolipidemic effects of lovastatin and

clofibrate alone and in combination in patients with type III hyperlipoproteinaemia. Metabolism 1990; 39: 403–409

104 Kushwaha RS, Hazzard WR, Gagne C, Chait A, Albers AA. Type III hyperlipoproteinaemia: Paradoxical hypolipidaemic effects of oestrogen. Ann Intern Med 1977; 87: 517–525

105 Goldstein JL, Hazzard WR, Schrott HG, et al. Hyperlipidaemia in coronary heart disease. Lipid levels in 500 survivors of myocardial infarction. J Clin Invest 1973; 52: 1533–1543

106 Manninen V, Elo MO, Frick MH, et al. Lipid alterations and decline in the incidence of coronary heart disease in the Helsinki Heart Study. J Am Med Assoc 1988; 250: 641–651

107 Sullivan DR, Sanders TAB, Trayner IN, Thompson GR. Paradoxical elevation of LDL and apoprotein B levels in hypertriglyceridaemic patients and normal subjects ingesting fish oil. Atherosclerosis 1986; 61: 129–134

108 Illingworth DR, Connor WE, Hatcher LF, Harris WS. Hypolipidaemic effects of n-3 fatty acids in primary hyperlipoproteinaemia. J Intern Med 1989; 225 (suppl.1): 91–97

109 Simmons LA, Hickie JB, Ballasubramian S. On the effects of omega-3 fatty acids (MaxEPA) on plasma lipids and lipoproteins in patients with hyperlipidaemia. Atherosclerosis 1985; 54: 75–88

110 Harris WS, Connor WE, Alam N, Illingworth DR. Reduction of postprandial triglyceridaemia in humans by dietary n-3 fatty acids. J Lipid Res 1988; 29: 1451–1460

111 Borkman M, Chisholm DJ, Furler SM. Effects of fish oil supplementation on glucose and lipid metabolism in NIDDM. Diabetes 1989; 38: 1314–1319

112 Leaf DA, Connor WE, Illingworth DR, Bacon SP, Sexton G. The hypolipidaemic effects of gemfibrozil in type V hyperlipidaemia. J Am Med Assoc 1989; 262: 3154–3160

113 Kesaniemi YA, Grundy SM. Influence of gemfibrozil and clofibrate on metabolism of cholesterol and plasma triglycerides in man. J Am Med Assoc 1984; 251: 2241–2246

114 Carlson LA, Olsson AG, Ballantyne D. On the rise in low density and high density lipoproteins in response to the treatment of hypertriglyceridaemia in type IV and V hyperlipoproteinaemias. Atherosclerosis 1977; 26: 603–609

115 Brunzell JD, Beirman EL. Chylomicronaemia syndrome: interaction of genetic and acquired hypertriglyceridaemia. Med Clin North Am 1982; 66: 455–468

116 Gordon DJ, Rifkind BM. High density lipoprotein. The clinical implications of recent studies. N Engl J Med 1989; 321: 1311–1316

117 Vega GL, Grundy SM. Comparison of lovastatin and gemfibrozil in normolipidaemic patients with hypoalphalipoproteinaemia. J Am Med Assoc 1989; 262: 3148–3153

118 Vega CL, Grundy SM. Lovastatin therapy in nephrotic hyperlipidaemia. Effects on lipoprotein metabolism. Kidney Int 1988; 33: 1160–1168

119 Golper TA, Illingworth DR, Morris CD, Bennett WM. Lovastatin in the treatment of multifactorial hyperlipidaemia associated with proteinuria. Am J Kidney Dis 1989; 13: 312–320

120 Rabelink AJ, Hene RJ, Erkelens DW, Joles JA, Koomans HA. Effects of simvastatin and cholestyramine on lipoprotein profile in hyperlipidaemia of nephrotic syndrome. Lancet 1988; 2: 1335–1338

121 Garg A, Grundy SM. Lovastatin for lowering cholesterol in noninsulin dependent diabetes mellitus. N Engl J Med 1988; 318: 81–86

122 Yoshino G, Kazumi T, Iwai M, et al. Long term treatment of hypercholesterolaemic noninsulin dependent diabetics with pravastatin (CS514). Atherosclerosis 1989; 75: 67–72

123 Betteridge DJ. Diabetes lipoprotein metabolism and atherosclerosis. Br Med Bull 1989; 45: 285–311
124 Newman WP III, Freedman DS, Voors AW, et al. Relation of serum lipoprotein levels and systolic blood pressure to early atherosclerosis: the Bogalusa Heart Study. N Engl J Med 1986; 314: 138–144
125 Garcia RE, Moodie DS. Routine cholesterol surveillance in childhood. Pediatrics 1989; 84: 751–755
126 American Heart Association position statement: Diagnosis and treatment of primary hyperlipidaemia in childhood. Circulation 1986; 74: 1181A–1188A

British Medical Bulletin (1990) Vol. 46, No. 4, pp. 1059–1074
© The British Council 1990

Policies for the prevention of coronary heart disease through cholesterol-lowering

B M Rifkind
*Lipid Metabolism-Atherogenesis Branch National Heart, Lung and Blood Institute,
National Institutes of Health, Bethesda, Maryland, USA*

Policies for coronary heart disease prevention through cholesterol-lowering have been advocated by groups in the UK, the US and Continental Europe. Such policies are based on studies of the atherosclerotic plaque, experimental animals, genetic disorders, and lipoprotein metabolism, and from different epidemiological studies and clinical trials. Surveys of physicians and the public indicate a growing awareness of the importance of cholesterol.

Expert groups advocate mutually compatible high-risk and population-based strategies. Selective or whole adult population screening is being discussed as is whether it should be done exclusively in the doctor's office or should also include mass public screening. Improved standards for public screening are necessary. The *US National Cholesterol Education Program* has developed or is developing reports on: (1) the detection, evaluation and treatment of high blood cholesterol in (a) adults and (b) children and adolescents; and on (2) standardization of cholesterol measurement and (3) population-based strategies for controlling cholesterol.

Although statements advocating and describing policies for the prevention of coronary heart disease (CHD) by means of cholesterol-lowering have appeared quite frequently in the UK and elsewhere for many years it is only recently that the emergence of a sufficient consensus in many countries is leading to their

0007–1420/90/0046–1059/$10.00

implementation. In the US the major medical professional organizations, under the aegis of the *National Heart, Lung and Blood
Institute's National Cholesterol Education Program* (NCEP) have
committed themselves to the goal of cholesterol-lowering.[1]
Through the NCEP and their individual efforts many steps are
being taken towards implementing cholesterol-lowering policies.
The momentum in the United Kingdom also seems to be gather-
ing. For example, the *Working Group on Coronary Disease of the
British Cardiac Society* recently specified both population-based
and high risk cholesterol-lowering strategies.[2] However, as
recently noted in the *26th Report of the Committee of Public
Accounts*, the implementation of such strategies seems to be lag-
ging behind.[3]

Any set of policies for the prevention or treatment of disease
must have a sufficiently compelling scientific foundation. This is
especially so in the area of CHD given the scale of the public
health and clinical problem. In the US heart and vascular diseases,
despite considerable declines in their mortality, continue to
account for the largest number of deaths of any cause group being
accountable for an estimated 979 000 deaths in 1986 or 46.6% of
all deaths.[4] Three-quarters of these deaths were atherosclerotic
related—524 000 being due to CHD. Men and women were both
affected; there being 290 000 deaths from CHD in men and
252 000 coronary deaths in women. The morbid consequences of
CHD should not be overlooked. About 16% of the 34 000 000
hospitalizations in the US in 1986 was due to cardiovascular diseases, more than for any other diagnostic category. In that year
there were 3.3 million hospitalizations and 26 million hospital days
because of atherosclerotic-related diseases, mostly due to CHD.
Various yardsticks from the UK tell the same story. The largest
single cause of adult male and female deaths in the UK is CHD
and more than 180 000 UK adults die of CHD annually—more
than from cancer and road accidents combined. In 1985 the estimated direct cost of CHD to the National Health Service was
389.9 million pounds and in 1984 there were 178 870 hospital in-
patient cases of CHD in England alone. In fact the UK now has
one of the highest rates of CHD in the world.

THE SCIENCE BASE

The evidence relating plasma cholesterol levels to atherosclerosis
and CHD is derived from many different approaches and one

should view it in its entirety in building an integrated picture rather than focusing on any one approach or study with its particular strengths and weaknesses. It can be classified as follows:

Atherosclerotic plaque

Cholesterol was first suspected to be involved in atherosclerosis after it was noted to be a prominent component of the early or the advanced atherosclerotic lesion. Contemporary studies are gradually defining the role of low density lipoprotein (LDL) and other lipoproteins in atherogenesis and the mechanisms underlying the formation of the early and more mature lesions.[5,6] Early atherosclerotic lesions are characterized by the accumulation of foam cells in the intima, which are mainly macrophages derived from blood monocytes. The macrophages enter the intima as a result of the release of chemotactic factors released by injured endothelial cells and then take up lipoproteins by means of surface receptors. It is believed that LDL is first modified by processes such as oxidation in order to be recognized by the receptors. Arterial smooth muscle cells migrate from the media to the intima and may also be transformed into foam cells. Some smooth muscle cells secrete large quantities of the connective tissue matrix, collagen and proteoglycans.

Animal studies

Lesions resembling human atherosclerosis have been produced in various animal species—including avian animal models (such as pigeons, quail and chickens), mammalian nonprimate models (such as rabbits and hares, swine and dogs) and, most importantly, many nonhuman primates—by inducing hypercholesterolemia usually by increasing the dietary saturated fat or cholesterol intake of the animal.[7] In some species this leads to myocardial ischemia and even infarction. Spontaneous forms of hypercholesterolemia in animals also occurs—for example in the *Watanabe heritable hyperlipidemic rabbit*, which develops lesions that closely resemble those found in cholesterol-fed rabbits and which also lead to severe atherosclerosis and myocardial ischemia.[8] The lipoprotein abnormalities and the underlying defect in the gene coding for the LDL receptor in the Watanabe rabbit also resemble those that occur in human homozygous familial hypercholesterolemia.

Clinical genetic disorders

Different human genetic diseases, which have raised concentrations of cholesterol-rich lipoproteins resulting from specific mutations in different genes, are characterized by severe atherosclerosis and often by the occurrence of CHD at a relatively young age. Familial hypercholesterolemia is the best known—either in its very common heterozygous variety or in its rarer but much more severe homozygous form.[9] Other examples are familial dysbetalipoproteinemia and familial combined hyperlipidemia.

Lipoprotein metabolism[6]

Over the past few decades the structure and metabolic functions of the various plasma lipoproteins have been gradually defined as have the factors controlling their metabolism especially dietary saturated fat and cholesterol. Each lipoprotein class is thought to play a distinctive role in cholesterol transport and metabolism and, consequently in the development of atherosclerosis. Low density lipoproteins appear to be atherogenic while high density lipoproteins (HDL) may provide protection against atherosclerosis.

Epidemiological studies[10]

Many different types of epidemiological studies have established the plasma cholesterol levels, specifically LDL cholesterol levels, as one of the several major risk factors for CHD in men and women, including the elderly. These include case-control studies, comparisons of populations with low and high rates of CHD, migrant studies, and international studies of diet, atherosclerosis, lipid levels and CHD. Especially prominent are the findings of many within-population prospective observational studies such as *Framingham*,[11] the *Multiple Risk Factor Intervention Trial* screenees,[12] and the *British Regional Heart Study*.[13] Such studies have consistently shown the cholesterol level to be a powerful and independent predictor of CHD so that each 1% rise in cholesterol is associated with an approximate 2% increase in CHD risk. The level of HDL has also been shown to be independently but inversely related to CHD in both sexes including the elderly.

Clinical trials

Although much evidence has been available in each of the above categories for many years and has been considerably amplified recently, physicians have been reluctant to embark on a comprehensive policy of cholesterol-lowering until clinical trial evidence was available to confirm that it was indeed beneficial and, especially given the long-term nature of such treatment, that it was sufficiently safe.

The results of several recent studies, especially when taken in conjunction with the findings of a large number of previous trials have provided compelling evidence that cholesterol-lowering reduces both coronary morbidity and mortality. Thus the *Coronary Primary Prevention Trial* (CPPT) showed that cholestyramine resin-induced cholesterol-lowering in hypercholesterolemic middle-aged men produced an average 19% reduction in the rate of definite fatal and/or nonfatal myocardial infarction over 7 years; men who took the full dose of drug and obtained greater than 25% cholesterol-lowering reduced their risk of CHD by half.[14,15] In the *Helsinki Heart Study* use of the fibric acid derivative gemfibrozil, which produced a moderate reduction in LDL and a moderate increase in HDL reduced fatal and nonfatal myocardial infarction by 34%.[16] A follow-up of patients with a prior history of myocardial infarction who previously received nicotinic acid for 5 years in the *Coronary Drug Project* showed them to have a reduced total and cardiovascular mortality when compared with the control group.[17] Recently, both the *Stockholm Ischemic Heart Disease Study* and the *Oslo Diet/Smoking Study* have reported beneficial outcomes including reduced total mortality.[18,19] Aggregate analysis of these and many other clinical trials suggest that cholesterol-lowering whether induced by diet or various drugs, either in the context of primary or secondary prevention, reduces fatal and nonfatal myocardial infarction.[20]

These results are supported by the findings from several angiographic studies, especially the *Cholesterol-Lowering Atherosclerosis Study* (CLAS) and the *Familial Atherosclerosis Treatment Study* (FATS), in which pronounced lipid change results in decreased progression and even in regression of atherosclerotic lesions in the coronary arteries.[21,22]

Introduction of HMG Co-A reductase inhibitors

The recent introduction of a new class of cholesterol-lowering drugs has undoubtedly contributed to the greater attention now being given to cholesterol-lowering. These agents which, unlike the resins and nicotinic acid, are easy to take and produce substantial reductions in total and LDL cholesterol, have been acceptably safe over the short term.[23] Alone or in combination with other drugs together with diet, it is now possible to reduce cholesterol levels to the desired range in most hypercholesterolemic individuals. However, the long-term safety of these agents will only be known after sufficient experience of their use has been obtained.

Public and physician perspective

Attitudes and practices of physicians regarding elevated plasma cholesterol levels have been assessed by National Heart, Lung, and Blood Institute-sponsored national telephone surveys of practising US physicians in 1983 and 1986.[24] The 1983 survey was conducted just before the release of the results of the CPPT. In 1986, 64% of physicians thought that reducing high blood cholesterol levels would have a large effect on heart disease, up considerably from 39% in 1983. Whereas in 1983 physicians attributed considerably less preventive value to reducing the cholesterol levels than to reducing blood pressure or smoking, this disparity was substantially smaller in 1986. The median range of blood cholesterol on which diet therapy was initiated was 6.21 to 6.70 mmol/l (240–259 mg/dl) in 1986, down from 6.72 to 7.21 mmol/l (260–279 mg/dl) in 1983; the median for drug therapy was 7.76–8.25 mmol/l (300–319 mg/dl) in 1986 and 8.7 to 9.28 mmol/l (340–359 mg/dl) in 1983. In 1986 87% of physicians surveyed felt that the medical evidence warranted recommended treatment levels set forth in the *1984 National Institutes of Health Consensus Conference on Lowering Blood Cholesterol.*[25] These changes indicate that by 1986, physicians were more convinced of the benefit of lowering high blood cholesterol levels and were treating patients accordingly. A further survey is being conducted and the results will be available shortly. It is likely that the continued emphasis on cholesterol-lowering will show a maintenance of these trends. Data from comparable public surveys also show gains in public awareness in action relating to high blood cholesterol levels.[26] Thus the percentage of adults who believe that reducing high blood cholesterol levels would have a large effect on heart disease increased from 64% in 1983 to 72% in 1986, so that the importance attached to reducing high blood cholesterol levels approached that attributed to reducing smoking and high blood pressure. In 1986, 46% of adults reported that they had their cholesterol levels checked compared with 35% in 1983. In both years diet changes were most frequently chosen (over 60%) as ways to control their blood cholesterol level; reducing dietary fat was believed to be as important as reducing dietary cholesterol. By 1986, 23% of adults reported that they had made dietary changes specifically to lower their blood cholesterol levels, up from 14% in 1983.

Similarly, from the UK it has been reported that whereas in 1982, 6% of those questioned in a public opinion poll mentioned high cholesterol levels as a cause of heart attacks, in 1989 70% believed that it had been proved that cholesterol-rich foods were bad for the heart.

Transatlantic policy statements[27]

The various scientific advances have been followed by appraisal of clinical and public health aspects of cholesterol and the lipid hypothesis and a large number of reports have appeared from countries in which CHD is a major problem. Almost without exception such reports recommend cholesterol-lowering strategies similar to the recommendations of the forty or so reports already in place summarized by Truswell in 1983.[28] In 1987 the European Atherosclerosis Society (EAS), with experts from 19 European countries including the United Kingdom, published its conclusions and recommendations on the *Strategies for the Prevention of Coronary Heart Disease*.[29] Also in 1987, the *British Cardiac Society (BCS) Working Group on Coronary Heart Disease Prevention* published its report.[2] These two policy statements joined that of the *US Consensus Conference Panel on Lowering Blood Cholesterol* (USCC) that appeared in 1985 in reflecting the viewpoint of much of the developed world on CHD prevention. The striking feature of these reports is the extent to which they agree on many issues which hitherto have been controversial. The Lancet noted the 'considerable agreement' between the EAS and BCS reports which reflected a 'medical consensus both in the UK and across the whole of Europe'.[30] In a leading article dealing with the BCS report, the British Medical Journal expressed the opinion that it could 'herald the start of a planned national attack on ischemic heart disease'.[31]

Each of these expert groups address the scientific evidence underlying the lipid hypothesis to a greater or lesser extent. Both the USCC and EAS unequivocally state that cholesterol plays a causal role in CHD. The concordant evidence advanced by many genetic, experimental, epidemiologic, and clinical trial research endeavours over the years are cited by these panels. The BCS report adopts a more conservative approach and states that total serum cholesterol has a continuous and curvilinear relationship to CHD risk.

The question of whether reduction of cholesterol levels help prevent CHD is answered affirmatively by the USCC and EAS, a review supported by the subsequent findings of the Stockholm, Oslo, Helsinki, CLAS and FATS trials.[16,18,19,21,22] Each of the three expert panels recommend that attempts should be made to reduce the blood cholesterol levels of the general population (Table 1), the so-called population-based approach to reducing CHD risk.

The three reports recommend high-risk as well as mass population approaches to the reduction of cholesterol and the management of other risk factors. While it is important to identify and treat all at high risk it is only through a population approach which seeks to promote health related behavioural changes in the whole population that the numerically much larger number of individuals at moderate risk and even low risk and who account for the largest absolute number of CHD cases can be reached. The high risk strategy calls for identifying those individuals above the 75th percentile for cholesterol, referring them to a physician and initiating treatment by diet, occasionally supplemented by drug therapy if necessary. Such individuals account for about 40–50% of all CHD cases in the population. However even if, in theory, all cases of CHD were eliminated in this group through such a strategy the 50–60% of all CHD that occurs in the 75% of the population with

Table 1 Views of expert panels

	US Consensus Conference 1985	European Atherosclerosis 1987	British Cardiac Society 1987
Is there a causal relationship between blood cholesterol levels and CHD?	Yes	Yes	A continuous and curvilinear relationship
Will reduction of blood cholesterol levels help help prevent CHD?	Yes	Yes	Results encouraging
Should an attempt be made to reduce blood cholesterol levels in the general population?	Yes	Yes	Yes

cholesterol levels below the 75th percentile would be neglected, indicating the need for a concomitant population-based strategy. It is important to emphasize that high risk or population based strategies are not mutually exclusive. Prevention of CHD will long continue to require both approaches and there is little room for competition between the two.[32] In fact as the EAS indicates, 'there is reason to believe that implementation of each strategy may enhance the effectiveness of the other. The agreement on this issue is best summed up by the BCS namely 'the population and high risk approaches are complementary and should be implemented simultaneously'.

US National Cholesterol Education Program (NCEP)[1]

Immediately after the announcement of the CPPT the NHLBI began to develop plans to ensure the widest dissemination of the CPPT results. This included the initiation of the NCEP which was, in general terms, modelled on the previously established *National High Blood Pressure Education Program (NHBPEP)*. An added impetus was received from the USCC; based on the review of the scientific evidence the consensus panel made a number of specific recommendations for further actions by the public, the health professional community, and government health agencies. In particular the panel recommended that,

> 'new and expanded programs be planned and initiated soon to edu-
> cate physicians, other health professionals, and the public to the
> significance of elevated blood cholesterol and the importance of
> treating it. We recommend that the National Heart, Lung, and
> Blood Institute provide the focus for development of plans for the
> NCEP that would enlist participation by contributions from inter-
> ested organizations at national, state, and local levels.'

The institute proceeded to implement the NCEP along these lines. Its overall objective is to reduce CHD morbidity and mortality related to elevated blood cholesterol by developing a national education effort and by stimulating extensive cooperation and coordination among responsible government agencies and interested public organizations.[1] As in the case of NHBPEP, the NCEP operates as a partnership. The Coordinating Committee was established to ensure effective mobilization of the resources and energies of all interested organizations. Its membership comprising representation from over 35 participating organizations including the American Medical Association, the American College of Cardiology, the American Heart Association, and the American Dietetic Association. This broad participation is indicative of the widespread agreement within US medicine that national policies on cholesterol-lowering are needed, though this is not to say that each of the participating organizations agrees on every question involving cholesterol.

Under the aegis of the NCEP and its coordinating committee four panels were established to address several important issues.

Laboratory standardization panel[33]

Measurement of blood cholesterol is clearly a prerequisite to the diagnosis and management of high blood cholesterol. The proficiency testing surveys of the College of American Pathologists have confirmed the general impression that there is considerable variation in the accuracy of cholesterol measurement.[33] The Laboratory Standardization Panel was established to examine this and related problems. Its membership included representatives from the major clinical chemistry organizations in the US, scientific equipment manufacturers and several US government agencies involved in various aspects of the laboratory measurement of cholesterol. It was charged with making recommendations towards improving the accuracy and precision of cholesterol measurement.

In its initial report the Laboratory Standardization Panel concluded that the current state of reliability of cholesterol measurements made in the US suggests that considerable inaccuracy in cholesterol testing exists. They concluded that obtaining reliable cholesterol values was quite feasible and set current and future standards for the accuracy and precision of measurement. They suggested that the key factor for obtaining accurate measurement was the use of reference materials with accurate target values for calibration and monitoring of the analytical process.

A special concern of the panel was the recent introduction of various devices for rapidly measuring cholesterol using capillary blood specimens. These instruments which have considerable promise in providing cheap, rapid, and near-painless total and HDL cholesterol measurement showed some initial shortcomings in their ability to measure cholesterol accurately but with improvements and appropriate attention to handling procedures and to their standardization they can provide results sufficiently accurate and precise for screening purposes.

The Laboratory Standardization Panel also examined the use,

by US laboratories of cutpoints (or reference values) for plasma cholesterol levels. Instead of the wide and often unrealistic range of values used they recommended that all laboratories adopt the cutpoints recommended by the Adult Treatment Panel of the NCEP (see below). Clearly implementation of the high risk strategy is greatly dependent on laboratories making this change.

Evaluation of high-risk subjects involves other lipid measurements; the Laboratory Standardization Panel has been succeeded by the Working Group on Lipoprotein Measurement which is considering the requirements of accurate and precise triglyceride, HDL-cholesterol and LDL-cholesterol measurement.

Population-strategies panel

The NCEP has also convened the Expert Panel on Population Strategies for Blood Cholesterol Reduction. Its report has just appeared and reviews the science-base for a population-based strategy, and the feasibility of achieving cholesterol-lowering through such a strategy and examines in detail the various societal steps and changes required to implement the strategy.

The USCC has already recommended that all Americans (except children under two years of age) be advised to adopt a diet that: reduces total dietary fat intake from the current level of about 40% of total calories to 30% of total calories; reduces saturated fat intake to less than 10% of total calories; increases polyunsaturated fat intake up to but no more than 10% of total calories; and reduces the daily cholesterol intake to 250–300 mg or less.

These recommendations are similar to those previously made by the American Heart Association and the Intersociety Commission for Heart Disease Resources.[34,35] They have since been repeated in the US by the Surgeon General's report on Nutrition and Health, the USDA Dietary Guidelines and by the Food and Nutrition Research Board of the National Research Council.[36–38] The BCS report endorsed the recommendations of the Committee on Medical Aspects of Food Policy[39] that the consumption of fat in the United Kingdom should be reduced so that no more than 35% of food energy is derived from total fat and 15% of food energy from saturated fatty acids. An increase to 0.45 in the ratio of polyunsaturated fatty acids to saturated fatty acids was also recommended. More stringent total and saturated fat restrictions were recommended for individuals at special risk of CHD.

The USCC report also focussed on the means of implementing

a population-based strategy. In addition to recommending new and expanded programs directed to physicians and other health professionals and the public a number of other specific recommendations were made. The food industry was encouraged to continue and intensify efforts to develop and market foods that would make it easier for individuals to adhere to the recommended diets and school food services and restaurants were urged to serve meals consistent with these dietary recommendations.

It was advocated that food labels should include the specific source(s) of fat, content of total fat, saturated and polyunsaturated fat, and cholesterol content as well as other nutritional information and that the public should be educated on how to use this information to achieve the dietary aims.

All physicians were encouraged to include whenever possible a blood cholesterol measurement on every adult patient when that patient is first seen. The NCEP advises each adult American to 'know your cholesterol level'.

Public screening for cholesterol

Irrespective of whether the complete adult population should be the target of screening or whether it should be confined to certain higher risk groups widespread screening of cholesterol levels is required. This is leading to consideration of the best approaches for cholesterol screening. There is good agreement that screening for high blood cholesterol is best done in the context of the physician's office (opportunistic screening) but consideration has also been given to mass or so-called public screening. Public screening has the possibility of detecting large numbers of individuals with high blood cholesterol levels in addition to those detected in the physicians' office. Unfortunately however the reliability of cholesterol measurements, and the education and follow-up in public screening programs are often inadequate. The results of current research and screening programs were presented at the NHLBI *Workshop on Public Screening for High Blood Cholesterol* in October 1988 which developed guidelines for screening which were subsequently adopted by the NCEP.[40] The workshop participants suggested methods that can make public screening more effective in detecting high blood cholesterol levels in individuals who might otherwise not be identified in the health care system and that could ensure follow-up of appropriate cases and public education about cholesterol. Their recommendations emphasize that public screening must meet customary standards for recruitment, reliable measurement of cholesterol level, appropriate information, staff training and referral. Concerns have been expressed by the Inspector General of the Department of Health and Human Services who noted that cholesterol screening is prevalent and growing; individuals with varied experience are entering the screening business; problems exist with the screening environment, providing staff for sample collection, counselling, and referral, and infection control; and that state and federal regulation is minimal.[41] The Inspector General recommended discouragement of public cholesterol screening that did not meet the NCEP guidelines.

Expert Panel on Detection, Evaluation, and Treatment of High Blood Cholesterol in Adults[42] (The Adult Treatment Panel)

This panel dealt with the high-risk strategy and came up with a series of recommendations for the identification and management of high blood cholesterol. It amplified and in some cases modified the USCC recommendations as follows:

● The total blood cholesterol level should be the basis for initial patient classification and all blood cholesterol levels above 5.2 mmol/l (200 mg/dl) should be confirmed by repeat measurements with the average used to apply clinical decisions

● Other CHD risk factors should be taken into account in selecting appropriate follow-up measures for patients for borderline high cholesterol levels

● All patients with a level of 6.2 mmol/l (240 mg/dl) and above, which is classified as high blood cholesterol, should receive a lipoprotein analysis. Patients with borderline high blood cholesterol levels (5.2–6.1 mmol/l or 200–239 mg/dl) who in addition have definite CHD or two other CHD risk factors should also have a lipoprotein analysis performed

● The LDL cholesterol level is the basis for decisions about initiating dietary and drug therapy

● Patients with LDL cholesterol levels of 4.1 mmol/l (160 mg/dl) or greater are considered at high risk for CHD. These patients should be given cholesterol-lowering treatments

● Patients with borderline high risk LDL cholesterol levels (3.4–4.0 mmol/l or 130–159 mg/dl) should also be treated to lower their cholesterol if they have definite CHD or two other CHD risk factors.

Panel on Children and Adolescents

This panel is considering the question of cholesterol-lowering in children, a somewhat more controversial and difficult area. It includes issues such as whether there should be universal or selective screening of cholesterol levels in childhood and the impact on growth and development of dietary-induced cholesterol lowering. There is growing agreement that, at the least, dietary saturated fat and cholesterol intake should be curtailed in childhood; both the American Heart Association and the Committee on Nutrition of the American Academy of Pediatrics have issued rather similar recommendations with the only real difference between the two sets of recommendations being the absence of quantitative targets in the statement issued by the American Academy of Pediatrics.[38]

REFERENCES

1 Lenfant C. A new challenge for America: The National Cholesterol Education Program. Circulation 1986; 73: 855–856
2 British Cardiac Society. Report of British Cardiac Society working group on coronary disease prevention. London: British Cardiac Society, 1987
3 Committee of Public Accounts 26th report. Coronary Heart Disease. London: HMSO, 1989
4 National Center for Health Statistics. Annual summary of births, marriages, divorces and deaths for August, United States, 1986. Monthly Vital Statistics Report, Vol 35, No 13, August 1987
5 Ross R. The pathogenesis of atherosclerosis. N Engl J Med 1986; 314: 488–500
6 Babiak J, Rudel LL. Lipoproteins and atherosclerosis. Clin Endocrinol Metab 1987; 1: 515–550
7 Vesselinovich D. Animal models and the study of atherosclerosis. Arch Pathol Lab Med 1988; 112: 1011–1017
8 Hatanaka K, Ito T, Shiomi M, Yamamoto A, Watanabe Y. Ischemic heart disease in the WHHL rabbit: A model for myocardiol injury in genetically hyperlipidemic animals. Am Heart J 1987; 113: 280–288
9 Brown MS, Goldstein JL. A receptor-mediated pathway for cholesterol homeostasis. Science 1986; 232: 34–47
10 Stamler JS. Lectures on Preventive Cardiology. New York: Grune and Stratton, 1967
11 Castelli WP, Garrison RJ, Wilson PWF, Abbott RD, Kalousdian S, Kannel WB. Incidence of coronary heart disease and lipoprotein cholesterol levels. The Framingham Study. J Am Med Assoc 1986; 256: 2835–2838
12 Neaton JD, Kuller LH, Wentworth D, Borhani NO. Total and cardiovascular mortality in relation to cigarette smoking, serum cholesterol concentration and diastolic blood pressure among black and white males followed up for five years. Am Heart J 1984; 108: 759–770
13 Pocock SJ, Shaper AG, Phillips AN, Walker M, Whitehead TP. High density

lipoprotein cholesterol is not a major risk factor for ischaemic heart disease in British men. Br Med J 1986; i: 515–519

14 Lipid Research Clinics Program. The Lipid Research Clinics Coronary Primary Prevention Trial Results. I. Reduction in incidence of coronary heart disease. J Am Med Assoc 1984; 251: 351–364

15 Lipid Research Clinics Program. The Lipid Research Clinics Coronary Primary Prevention Trial Results II. The relationship of reduction in incidence of coronary heart disease to cholesterol lowering. J Am Med Assoc 1984; 251: 365–374

16 Frick MH, Elo O, Haapa K, et al. Helsinki Heart Study: Primary-Prevention Trial with gemfibrozil in middle-aged men with dyslipidemia. Safety of treatment, changes in risk factors, and incidence of coronary heart disease. N Engl J Med 1987; 317: 1237–1245

17 Canner PL, Berge KG, Wenger NK et al. Fifteen year mortality in coronary drug project patients: long-term benefit with niacin. J Am Coll Cardiol 1986; 8: 1245–55

18 Carlson LA, Rosenhamer G. Reduction in mortality in the Stockholm ischaemic heart disease secondary prevention study by combined treatment with clofibrate and nicotinic acid. Acta Med Scand 1988; 223: 405–418

19 Hjermann I, Holme I, Leren P. Oslo Study diet and antismoking trial. Results after 102 months. Am J Med 1986; 80(2A)7–11

20 Yusuf S, Wittes J, Friedman L. Overview of results of randomized clinical trials in heart disease II. Unstable angina, heart failure, primary prevention with aspirin and risk factor modification. J Am Med Assoc 1988; 260: 2259–2263

21 Blankenhorn DH, Nessim SA, Johnson RL, Sanmarco ME, Azen SP, Cashin-Hemphill L. Beneficial effects of combined colestipol-niacin therapy on coronary atherosclerosis and coronary venous bypass grafts. J Am Med Assoc 1987; 257: 3233–3240

22 Brown BG, Lin JT, Schaefer SM, Kaplan CA, Dodge HT, Albers JJ. Niacin or lovastatin combined with colestipol, regress coronary atherosclerosis and prevent clinical events in men with elevated apolipoprotein B. Circulation 1989; 80: II–266

23 Gordon DJ, Rifkind BM. 3-hydroxy-3-methylglutaryl coenzyme A (HMG-CoA) reductase inhibitors: A new class of cholesterol-lowering agents. Ann Intern Med 1987; 107: 759–761

24 Schucker BH, Wittes JT, Cutler JA et al. Change in physician perspective on cholesterol and coronary heart disease: Results from two national surveys. J Am Med Assoc 1987; 258: 3521–3526

25 Consensus Conference Statement on lowering blood cholesterol to prevent heart disease. J Am Med Assoc 1985; 253: 2080–2086

26 Schucker B, Bailey K, Heimbach JT et al. Change in public perspective on cholesterol and heart disease. Results from two national surveys. J Am Med Assoc 1987; 258: 3527–3531

27 Brook JG, Rifkind BM. Cholesterol and coronary heart disease prevention—A translantic consensus. Eur Heart J 1989; 10: 702–711

28 Truswell AS. The Development of Dietary Guidelines. Food Technology in Australia 1983; 35: 498–502

29 Study Group, European Atherosclerosis Society. Strategy for the prevention of coronary heart disease: A policy statement for the European Atherosclerosis Society. Eur Heart J 1987; 8: 77–88

30 Editorial, Prevention of coronary heart disease. Lancet 1987; i: 601–2

31 Hart JT. Coronary prevention in Britain: action at last? Br Med J 1987; 294: 725–6

32 Rose G. Sick Individuals and Sick Populations. Int J Epidemiol 1985; 14: 32–38

33 Current status of blood cholesterol measurement in clinical laboratories in the United States: A report from the Laboratory Standardization Panel of the National Cholesterol Education Program. Clin Chem 1988; 34: 193–201
34 American Heart Association. Diet and coronary heart disease. Dallas: American Heart Association, 1978
35 Intersociety Commission for Heart Disease Resources. Primary prevention of the atherosclerotic diseases. Circulation 1970; 42: A 55–95
36 The Surgeon General's Report on Nutrition and Health. Washington DC. U.S. Department of Health and Human Services. DHHS (PHS) Publication No. 88-50210, 1988
37 U.S. Department of Agriculture and U.S. Department of Health and Human Services 1985. Nutrition and your health; dietary guidelines for Americans. 2nd edn. Home and Garden Bulletin No. 232. Washington, DC: U.S. Government Printing Office, 1985
38 Committee on Diet and Health and Food and Nutrition Board. Diet and health. Washington, DC: National Academy Press, 1989
39 Committee on Medical Aspects of Food Policy, Department of Health and Social Security. Diet and cardiovascular disease. London: HMSO, 1984
40 Office of Inspector General, Department of Health and Human Services. Public cholesterol screening. OA1-05-89-01330, 1989
41 Recommendations regarding public screening for measuring blood cholesterol. Summary of a National Heart, Lung, and Blood Institute Workshop, October 1988. Arch Intern Med 1989; 149: 2650–2654
42 Report of the National Cholesterol Education Program Expert Panel on Detection, Evaluation, and Treatment of High Blood Cholesterol in Adults. Arch Intern Med 1988; 148: 36–69

British Medical Bulletin (1990) Vol. 46, No. 4, pp. 1075–1087

Cholesterol as a risk factor for coronary heart disease

H Tunstall-Pedoe
W C S Smith
Cardiovascular Epidemiology Unit, Ninewells Hospital and Medical School, Dundee, Scotland

The intense current interest in lipids as causal risk factors in coronary heart disease has encouraged a unifactorial, one-dimensional approach to coronary risk based on fixed total serum cholesterol or low density lipoprotein (LDL) cholesterol cut-points. Epidemiological evidence shows that coronary heart disease risk is multifactorial; the risk associated with a given lipid value is modulated overwhelmingly by the level or presence of other factors. While populations with high lipid values may be characterized as being at high risk, individual risk cannot be determined by isolated lipid measurements. To characterize individuals as at high or low risk by considering lipid values alone results in serious misclassification. Cut-points derived from middle-aged American men create anomalies when applied to different age and sex groups and to different populations. Clinical management of risk factors in individuals should involve the negotiation of a flexible and multi-dimensional individual regime, comprising all modifiable factors.

The latter 1980s saw a crescendo of medical and lay interest in cholesterol and other lipids as coronary risk factors. Follow-up of *Multiple Risk Factor Intervention Trial* (MRFIT) screenees[1] and the results of the *Lipid Research Clinics Coronary Primary Prevention Trial*[2] and the *Helsinki Heart Study*[3], as well as the awarding of a Nobel prize for work in cholesterol[4] were associated with the launching of the *National Cholesterol Education Program* by the National Heart Lung and Blood Institute (NHLBI) of the USA[5],

0007–1420/90/0046–1075/$10.00

and the issuing of policy statements by an *International Study Group of the European Atherosclerosis Society*[6,7]. Contemporaneous with these was the intensive marketing of the drugs which had figured in the two trials, plus the licensing and launch of new powerful lipid-lowering drugs.[8]

Follow-up of the MRFIT screenees, numbering 360 000[1] showed that, at least in middle-aged American men, and over the range of serum cholesterol found between their top and bottom deciles, there is no apparent threshold value of serum cholesterol at which risk of a coronary death begins to increase. Although the risk gradient steepens with increasing serum cholesterol, there is no lower level at which risk is completely flat. Should cholesterol associated coronary risk be reversible there is therefore no bottom limit, no point at which lowering cholesterol further would theoretically produce no dividend at all for coronary risk. Viewed in those terms therefore it can and is being argued that everyone might be expected to benefit from a lowering of serum cholesterol, even if the maximum benefit from a specified reduction would accrue to those with the highest levels.

The argument therefore is that coronary risk from serum total cholesterol is graded and continuous.[9] Evidence for reversibility of risk however, is weaker and less definite than that for risk itself. Nonetheless, two major randomized, placebo-controlled drug trials have now reported showing that coronary risk is, to an extent, reversible when serum cholesterol levels are lowered. The *Lipid Research Clinic Trial*[2] was the first to report. Although there was some controversy over the fact that the statistical test that was used to measure the significance of the result was single-tailed and not two-tailed, the congruence of the results for different coronary heart disease related end-points, and the fact that reduction of risk could be correlated with the degree of cholesterol lowering and with the compliance with drug dosage were convincing.[10] The *Helsinki Heart Study* confirmed these results and achieved an unequivocal result.[3,11] Neither study achieved a statistically significant reduction in coronary deaths, or in all causes mortality—they were not designed to do so. These two studies have therefore shown that some cholesterol associated risk is reversible. Although the cautious can argue that what was found applies only to middle-aged men satisfying the entry criteria for the two studies, and treated with the drugs concerned, many now consider that these trials were conclusive tests of cholesterol as a causal factor, and should be taken to mean (with the meta-analysis of other smaller and less conclusive studies) that the cholesterol risk gradient is reversible, and can be descended by lowering serum cholesterol by whatever means.

Unequivocal reversibility of risk of any of the other major coronary risk factors such as cigarette smoking and blood pressure[12] has not been similarly demonstrated by unifactorial randomized controlled trials. The sense of excitement about lowering coronary risk through cholesterol has been contributed to by lavish drug advertising and sponsorship of medical meetings and educational material. Commercial interests are in the business of creating a large market and exploiting it, and in doing so the literature that they sponsor often oversimplifies the epidemiological evidence on which the argument for therapy is based. The purpose of this paper is to seek to reestablish a balance. To do this it will concern itself solely with the interaction of major coronary risk factors and their importance in interpreting cholesterol associated risk. Because the epidemiological evidence of the interaction of total serum (or plasma) cholesterol with the other major risk factors is so much more extensive, the cholesterol fractions will not be discussed separately. This is not do deny their importance nor that of other risk factors such as age, sex, family history, diabetes and evidence of coronary heart disease.[13] However, the same principles of interaction will apply to them as to the total serum cholesterol.

MRFIT SCREENEES REVISITED

The follow-up of MRFIT screenees constitutes the largest follow-up study of major coronary risk factors and coronary heart disease mortality ever conducted. Its results are of importance in that the large number of end-point events has enabled smooth curves to be generated from real data rather than from idealized computer models, and that the effects of risk factor interaction can be looked at directly. Many previous studies have published results based on inadequate numbers of end-point events, which can produce anomalies and inconsistencies.[14] However, it must not be assumed that what is true of middle-aged American men screened in the early 1970s will be true of men of all age-groups everywhere and of women also. Not only does the event rate vary in different studies, but the relative contribution of the major risk factors to overall risk also varies. This was first found in the *Seven Countries Study* where it was shown that cigarette smoking seemed to be much more of a coronary risk factor in some populations than in others.[15]

The results of the MRFIT screenees have often been presented as showing a smooth gradient of risk for serum cholesterol,[5,6] derived from Figure 3 in Martin et al.[1] It is argued that the gradient of risk accelerates from a value of 200 mg/dl (5.2 mmol/l) and that the risk becomes serious beyond 240 mg/dl (6.2 mmol/l) or 250 mg/ dl (6.5 mmol/l) so that those above these latter cut-points should be considered as at high relative risk. This is true of total serum cholesterol considered in isolation. However, if the graphs of gradient of risk by percentile of cholesterol and of diastolic blood pressure are superimposed as in Figure 4 in Martin et al.[1] it is apparent that the gradient of risk and the risk multiple between highest and lower percentiles is almost the same, as the curves for blood pressure and serum cholesterol lie almost on top of each other. In the MRFIT screenees diastolic blood pressure is a risk factor for coronary death that is as powerful as is total serum cholesterol.

Further evidence of the importance of the other risk factors in this study is shown in a more detailed publication of this group which may not have achieved the prominence it deserved because the data are largely presented in multiway tables.[16] Two age groups, five quintiles of blood pressure, five quintiles of serum cholesterol and two smoking groups (smokers and non-smokers) result in 100 cells, which are difficult to show graphically. Figure 1 shows what happens when blood pressure is ignored, but the effect of age and smoking on serum cholesterol are plotted. For any given

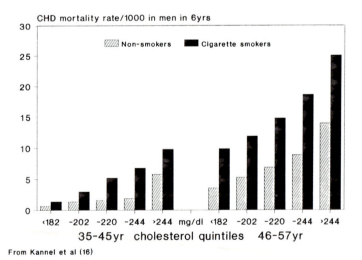

From Kannel et al (16)

Fig. 1 Six year coronary heart disease mortality by quintile of serum cholesterol, age group and smoking group.

quintile (fifth) of serum cholesterol, coronary risk is much greater in smokers than in non-smokers. In addition the risk in the older age group, for any given level of risk factors, is higher than that in the younger one. The same interaction is true of cholesterol and blood pressure. We have already presented a sophisticated analysis of these data in the *British Medical Journal*[17] showing how the other risk factors modulate cholesterol associated (or attributable) risk. This is shown again in Figure 2. For each stratum of age, smoking and blood pressure the coronary death rate in the lowest quintile of serum cholesterol has been subtracted from that in the highest. This produces a cholesterol attributable risk, or measure of the risk carried by those with high rather than low cholesterol values. For the same differences in cholesterol readings the risk varies by a factor of fifteen or more as a result of age, smoking and blood pressure. It is therefore not reasonable to consider that everyone whose cholesterol is in the top quintile is at high coronary risk as many such men have low levels of other risk factors. Nor is high coronary risk confined to those with high cholesterol, as the effects of smoking and/or raised blood pressure can be just as serious, and the combination of moderate levels of all three is far worse than elevation of one only.

Paradoxically therefore, the results from the MRFIT screenees,

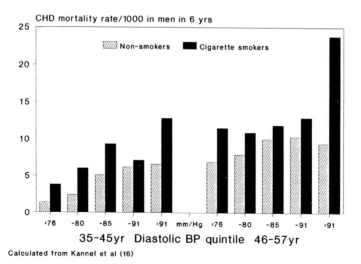

Calculated from Kannel et al (16)

Fig. 2 Additional (or attributable) risk of death from coronary heart disease at each level of age, cigarette smoking and blood pressure through having a high cholesterol (top quintile) rather than a low cholesterol (bottom quintile). For further explanation see text and Reference 17.

which have been used to promote a unidimensional, cholesterol-based approach to coronary risk assessment, provide the best possible evidence that this approach is looking at only a third or so of the problem.

CHOLESTEROL VALUES IN THE POPULATION

The mean values of serum cholesterol vary in different populations as was shown by many including the *Seven Countries Study*.[15] The most recent data have come from the *World Health Organization MONICA Study*[18], of trends and determinants of cardiovascular disease which has published cross-sectional data from the first risk factor survey.[19] Median total cholesterol in men aged 35–64 varied from 4.1 mmol/l to 6.4 mmol/l across fifty populations. Less well known is that there are changes in mean levels with age even during adult life, and that these are particularly marked in women. Data from Scottish men and women are shown in Figures 3 and 4 in the form of decremental frequency distributions or 'how-often-that-high' graphs.[20] The Y axis of these graphs shows the

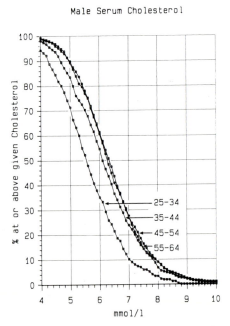

Male Serum Cholesterol

Fig. 3 How-often-that- high graph of serum cholesterol in men by ten year age groups 25–64 years.

Female Serum Cholesterol

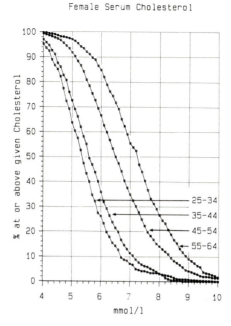

Fig. 4 How-often-that- high graph of serum cholesterol in women by ten year age groups 25–64 years.

percentage of the population with total cholesterol values at or above those shown on the X axis for each ten year age and sex group. It is apparent from these graphs that the proportion of the population affected by the use of fixed cut-points will vary very markedly with the mean, and that within populations it will vary by age and sex. Any cut-point in women will implicate a disproportionate number of women over the age of 45—far more than the equivalent cut-point in men. In our data about a third of the difference between men and women at these ages is accounted for by higher high-density liproprotein cholesterol.

THE NHLBI AND EUROPEAN ATHEROSCLEROSIS SOCIETY GUIDELINES

Our examination of the interaction of the coronary risk factors and the distribution of cholesterol by age and sex enables us to examine the two cholesterol management algorithms that have been published, one in the USA,[5] and one in Europe.[6] Both recommend mass testing (or opportunistic screening) of serum cholesterol, both

Table 1 Percentages exceeding management algorithm cut-points by age and sex in the Scottish population

Age group		25–34	35–44	45–54	55–64	Scottish population25–64 (1986)
5.2 mmol/l	Male	63	76	86	84	75
(200 mg/dl)	Female	57	66	88	95	75
6.2 mmol/l	Male	29	46	54	53	44
(240 mg/dl)	Female	20	29	60	79	45
6.5 mmol/l	Male	21	36	43	41	34
(250 mg/dl)	Female	14	21	50	71	37
7.8 mmol/l	Male	4	11	11	10	9
(300 mg/dl)	Female	3	4	17	31	13

use fixed cut-points to determine action, and both state that their cut-points should be used regardless of age and sex, and in different populations regardless of the values of cholesterol commonly present in those populations. Each of them recommends measurement of total serum cholesterol initially, but that cholesterol fractionation (more expensive and subject to more laboratory variation) should be carried out on everybody whose initial cholesterol reading exceeds a very common figure. Table 1 shows the percentage of the Scottish population exceeding the NHLBI and European guidelines. Between a third and a half of the adult population would be implicated on the basis that they were at high coronary risk. However, we have already seen that the latter is not necessarily true. Although both documents stress the importance of other risk factors, the two management algorithms differ in how they treat them. The NHLBI document does modify its management algorithm somewhat depending on the sex of the subject and the presence of other major coronary risk factors. The European document makes no distinction by age or sex in its basic algorithm, and its summary statement, that everyone with a cholesterol over 6.5 mmol/l should be under clinical care, and everyone over 7.8 mmol/l should be referred to a Lipid Clinic, is frequently repeated without qualification in drug literature, such as the Monthly Index of Medical Specialties (MIMS) circulated to British doctors. The proportion of post-menopausal women implicated as apparent from Table 1, and the naivety of such advice is clear from the previous discussion.

CALCULATION OF MULTIFACTORIAL RISK

Notwithstanding the importance of serum total cholesterol as a coronary risk factor, the cholesterol management algorithms produced by NHLBI and the European Atherosclerosis Society have not been universally accepted. In Britain the case for testing everyone's cholesterol has not been established. Whilst it is true that you cannot know what it is without measuring it, the interaction with other risk factors means that before deciding whether to measure it you can know how much, if at all, the result is going to matter.[17] Cholesterol does not satisfy some of the key criteria demanded for introduction of a screening test into clinical practice,[21] and a British consensus conference in 1989 recommended selective testing only,[22] as did the British Medical Association. Single measurements of cholesterol result in a significant degree of misclassification, because of within-person variability, and this misclassification is only slowly remedied by serial measurements.[23]

Because serum total cholesterol is a risk factor for coronary heart disease, and for very little else, (something that is easily forgotten in lipid clinics where rare metabolic abnormalities involving other lipids are seen), its relevance is virtually confined to its contribution to coronary risk. For this reason more of the energy currently vested in measuring cholesterol and its fractions needs to be directed to finding acceptable and usable coronary risk scoring systems which incorporate the other major modifiable factors, help clinicians to decide on priorities, and help patients to get their risk factor problems in proportion.

One approach to this is being developed in the Cardiovascular Epidemiology Unit in Dundee. Called the Dundee Risk-Score Risk-Rank calculator (Fig.5), it is really a circular slide rule for multiplying together the risks derived from cigarette smoking, blood pressure and serum cholesterol (either a measured value or a dummy value if it has not been measured). The readout tells the doctor and the patient either the patient's relative risk for a coronary event over five years based on these three factors, or alternatively what his rank (1 to 99) is compared with the general population, rank 1 being at the front of the queue for a coronary, and 99 being at the rear. The calculator is very rapid to use, identifies priority groups for management, and should be of educational value to the patient. It does not presume that cholesterol tests have been done on everybody, but can give an indication as to which patients, based on other risk factors, might merit cholesterol measurement.

The formula for this calculator is a multiple logistic function derived from follow-up of men in the United Kingdom Heart Disease Prevention Project.[24] The calculation of risk is as follows:

$$y = \frac{1}{1 + e^{-(a + b_1 x_1 + b_2 x_2 + b_3 x_3)}}$$

$y =$ risk of hard criteria coronary heart disease in 5 years for men free of CHD aged 40–59 at entry.

$a = -6.8684$
$b_1 = 0.010496$ $x_1 =$ systolic blood pressure
$b_2 = 0.36151$ $x_2 =$ cholesterol mmol/l
$b_3 = 1.00$ $x_3 =$ smoking code

never	$= 0$
ex-smoker, pipe or cigars	$= 0.14514$
1–10 cigarettes	$= 0.64950$
11–20 cigarettes	$= 0.98505$
> 20 cigarettes	$= 1.3530$

The advantage of such a formula and calculator is that it demonstrates the comparative importance of risk factors and their interaction. For example the difference between never smoking and smoking more than 20 cigarettes a day is a trebling of risk, and

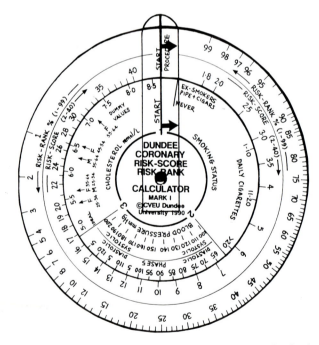

Fig. 5 The (draft) Dundee Risk-Score Risk-Rank Calculator (under development and copyright).

this is also true of a cholesterol of 8.5 mmol/l versus 5.0 mmol/l. A high cholesterol is therefore three times more serious in a heavy smoker than in a non-smoker, so that a multifactorial approach is justified both for risk assessment and for its management. Because three major coronary risk factors are being considered, it is not too surprising that use of the dummy cholesterol reading, instead of an actual reading, leads to only a third of men or women moving more than one decile of risk from where they would have been had the reading been available. As this calculation is based on the assumption that a single cholesterol reading is correct, the true misclassification may be smaller.

CONCLUSION

This paper has concentrated on the significance of the interaction of the three major risk factors and their implications for clinical practice. It has been written as an antidote to the current idea that cholesterol and coronary risk can be assessed and managed in one dimension by numbers—the cholesterol value.

We have shown that cholesterol is only one part of coronary risk and that it is dangerous to ignore the other contributors. Coronary risk should not be assessed in the laboratory, but at the bedside. Epidemiological statements on risk based on populations cannot be applied uncritically to individuals to determine patient management. Management algorithms for coronary risk in individuals should be based on overall coronary risk rather than on a single risk factor, as the relevance of any one is determined by the other risk factors. The risk factors interact multiplicatively. This means that cholesterol values should be considered as if they are marked out on a piece of elastic, rather than a ruler: the presence or level of other risk factors may or may not stretch out the significance of a particular cholesterol reading. Management should be flexible, and based on a negotiated contract with the patient, in which he or she agrees which areas of lifestyle should be given first priority for change.

None of this detracts from the importance of cholesterol in population terms as a primary reversible risk factor for coronary heart disease, nor the need for national nutrition policies to cope with the population cholesterol levels demonstrated in Figures 3 and 4, which are typical of Britain as a whole.[20] Recommendations for a healthy diet may (and this is still inadequately established) be reinforced by a high-risk strategy and by cholesterol testing, but they are not an essential prerequisite.

Note: The opinions expressed in this paper are those of the authors and **not** those of the Scottish Home and Health Department which funds the Cardiovascular Epidemiology Unit.

REFERENCES

1 Martin MJ, Hulley SB, Browner WS, Kuller LH, Wentworth D. Serum cholesterol, blood pressure and mortality: implications from a cohort of 361,662 men. Lancet 1986; ii: 933–6
2 Lipid Research Clinics Program. The Lipid Research Clinics Coronary Primary Prevention Trial results. I. Reduction in incidence of coronary heart disease. J Am Med Assoc 1984; 251: 351–64
3 Frick MH, Elo O, Haapa K, Heinonen OP, Heinsalmi P, Helo P et al. Helsinki Heart Study: primary-prevention trial with gemfibrozil in middle-aged men with dyslipidemia. N Engl J Med 1987; 317: 1237–45
4 Brown NS, Goldstein JL. A receptor mediated pathway for cholesterol homeostasis. Science 1986; 232: 34–37
5 US Cholesterol Education Campaign Expert Panel. Report of the national cholesterol education program expert panel on detection, evaluation and treatment of high blood cholesterol in adults. Arch Intern Med 1988; 148: 36–69
6 Study Group, European Atherosclerosis Society. Strategies for the prevention of coronary heart disease: a policy statement of the European Atherosclerosis Society. Eur Heart J 1987; 8: 77–88
7 Study Group, European Atherosclerosis Society. The recognition and management of hyperlipidaemia in adults: a policy statement of the European Atherosclerosis Society. Eur Heart J 1988; 9: 571–600
8 Maher VMG, Thompson GR. HMG CoA reductase inhibitors as lipid-lowering agents: five years experience with lovastatin and an appraisal of simvastatin and pravastatin. Q J Med 1990; 74: 165–175
9 Stamler J, Wentworth D, Neaton JD, for the MRFIT Research Group. Is relationship between serum cholesterol and risk of premature death from coronary heart disease continuous and graded? Findings in 356 222 primary screenees of the Multiple Risk Factor Intervention Trial (MRFIT). J Am Med Assoc 1986; 256: 2823–2828
10 Lipid Research Clinics Program. The Lipid Research Clinics Coronary Primary Prevention Trial Results II. The relationship of reduction in incidence of coronary heart disease to cholesterol lowering. J Am Med Assoc 1984; 251: 365–74
11 Manninen V, Elo O, Frick H, Haapa K, Heinonen OP, Heinsalmi P et al. Lipid alterations and decline in the incidence of coronary heart disease in the Helsinki Heart Study. J Am Med Assoc 1988; 260: 641–651
12 Medical Research Council Working Party. MRC trial of treatment of mild hypertension: principal results. Br Med J 1985; 291: 97–104
13 Gordon DJ, Rifkind BM. High-density lipoprotein—the clinical implications of recent studies. N Engl J Med 1989; 321: 1316
14 Phillips AN, Pocock SJ. Sample size requirements for prospective studies with examples for coronary heart disease. J Clin Epidemiol 1989; 42: 639–648
15 Keys A. Seven countries: a multivariate analysis of death and coronary heart disease. Cambridge Massachusetts: Harvard University Press, 1980
16 Kannel WB, Neaton JD, Wentworth D et al. Overall and coronary heart disease mortality rates in relation to major risk factors in 325,348 men screened for MRFIT. Am Heart J 1986; 112: 825–36

17 Tunstall-Pedoe H. Who is for cholesterol testing? Test selectively those who will benefit most. Br Med J 1989; 298: 1593–94
18 World Health Organization Principal Investigators (prepared by Tunstall-Pedoe H). The World Health Organization MONICA Project (monitoring trends and determinants in cardiovascular disease): a major international collaboration. J Clin Epidemiol 1988; 41: 105–114
19 WHO MONICA Project (prepared by Pajak A, Kuulasmaa K, Tuomilehto J, Kuokokoski E) Geographical variation in the major risk factors of coronary heart disease in men and women aged 35–64. World Health Stat Q 1988; 41: 115–137
20 Tunstall-Pedoe H, Smith WCS, Tavendale R. How-often-that-high Graphs of Serum Cholesterol. Findings from the Scottish Heart Health and Scottish MONICA Studies. Lancet 1989; 1: 540–2
21 Smith WCS, Kenicer MB, Maryon Davis A, Evans AE, Yarnell J. Blood cholesterol: is population screening warranted in the UK? Lancet 1989; i: 371–3
22 Consensus statement. Blood cholesterol measurement in the prevention of coronary heart disease. Sixth King's Fund Forum: King Edward's Hospital Fund for London. 1989
23 Jacobs D, Barrett-Connor E. Re-test reliability of plasma cholesterol and triglycerides. The Lipid Research Clinics Prevalence Study. Am J Epidemiol 1982; 116: 878–85
24 Rose G, Tunstall-Pedoe H, Heller RF. United Kingdom Heart Disease Prevention Project: incidence and mortality results. Lancet 1983; i: 1062–1065

Index

A

Age and cholesterol, 871
Alcohol intake, 1013–1014
Apo(a) gene: structure/function
    relationships and the possible link
    with thrombotic atheromatous
    disease: REES A, BISHOP A &
    MORGAN R, 873–890
Apolipoprotein
    genes, 928–934, 944–946
    regulatory sequences,
        identification of, 951–953
Arterial intima lipid accumulation,
    971–974
Atherosclerosis, 867
    pathology of, 960–985
    genetic susceptibility, 917–940

B

BARALLE F E see SIDOLI A
BISHOP A see REES A

C

Candidate genes, 924–925
CHAMBERLAIN J C & GALTON
    D J: Genetic susceptibility to
    atherosclerosis, 917–940
CHD = coronary heart disease
Cholestasis, 1018–1019
Cholesterol as a risk factor in coronary
    heart disease: TUNSTALL-
    PEDOE & SMITH W C S,
    1075–1086
Cholesterol screening, 1070–1071
Cholesterol-lowering for the
    prevention of CHD, 1059–1074
CLAS = Cholesterol-Lowering
    Atherosclerosis Study, 1063, 1066
Connective tissue proliferation and the
    arterial smooth muscle cell, 974–979
CPPT = Coronary Primary Prevention
    Trial, 1063–1064, 1075
CVD = cardiovascular diseases (see also
    CHD)

D

Diabetes mellitus, 1006–1012
Diagnostic tests, DNA based, 941–959
DNA based diagnostic tests:
    recombinant DNA and
    cardiovascular diseases risk factors:

SIDOLI A, GALLIANI S &
    BARALLE F E, 941–959
DURRINGTON P N: Secondary
    hyperlipidaemia, 1005–1024

E

Education, 871–872
Epidemiology, 1062–1063, 1075–1084

F

Familial lipase deficiencies, 987–989
FATS = Familial Atherosclerosis
    Treatment Study, 1063, 1066
FCH = familial combined
    hyperlipidaemia, 1001–1002
FH = familial hypercholesterolaemia,
    866, 935, 942, 945–946, 986,
    991–999
Framingham Study, 918, 931, 1062

G

GALLIANI S see SIDOLI A
GALTON D J see CHAMBERLAIN
    J C
Gene expression control, 946–948
Gene markers, 923–924
Genetic markers strategy, 943–944
Genetic polymorphisms, 928–937
Genetic susceptibility to
    atherosclerosis: CHAMBERLAIN
    J C & GALTON D J, 917–940
Genotype analysis by DNA based
    diagnostic tests, 943–953
Gout & hyperuricaemia, secondary
    hyperlipidaemia, 1019–1020
Growth factors, 976–978

H

Hardy-Weinberger equilibrium, 950
HDL = high density lipoproteins
Hepatocellular disease, 1019
HMG (3-hydroxy-3-methyl-glutaryl)
    CoA-reductase inhibitors,
    1063–1064
Hormones & effectors on LDL-
    receptors, 910–911
Hypercholesterolaemia, primary,
    991–999
Hyperlipidaemia
    children, 1051–1052

combined, drug therapy,
1046–1048
diagnosis of, 1026–1029
primary, 986–1004
mixed, 999–1002
secondary, 1005–1024
treatment of, 1025–1058
Hypertriglyceridaemia, 868, 987–991
Hyperuricaemia & gout, secondary
hyperlipidaemia, 1019–1020

I

ILLINGWORTH D R: Treatment of
hyperlipidaemia, 1025–1058
Insulin, 1006–1012
Introduction: Lipids and coronary
disease—resolved and unresolved
problems: OLIVER M F, 865–872

J

Japan study, 932–933

K

KNIGHT B L see SOUTAR A K

L

LDL = low density lipoproteins
LDL-receptor & its gene, 866,
891–916, 935
Lipoprotein lipase gene, 935–936
Liver disease, 1018–1019
Lp(a) = lipoprotein(a), 873–890
LPDS = lipoprotein deficient serum,
903–904, 908

M

MORGAN R see REES A
MRFIT = Multiple Risk Factor
Intervention Trial, 1075–1080
Multifactorial risk, calculation of,
1081–1084

N

NCEP = National Cholesterol
Education Program, 1029,
1059–1060, 1067–1070, 1075
Nephrotic syndrome, 1014–1015

O

Obesity, 1012–1013
OLIVER M F: Introduction: Lipids
and coronary disease—resolved
and unresolved problems,
865–872

P

Paediatric hyperlipidaemia therapy,
1051–1052
Pathology of atherosclerosis: WOOLF
N, 960–985
PDGF = platelet-derived growth
factor, 976–978
Phenotype analysis by DNA based
diagnostic test, 954–957
PIC value = polymorphism
information content value,
920–921
Plasminogen receptors, 883–885
Policies for the prevention of coronary
heart disease through cholesterol-
lowering: RIFKIND B M,
1059–1074
Prevention of CHD by cholesterol-
lowering, 1059–1074
Primary hyperlipidaemia:
THOMPSON G R, 986–1004
Prospective Basle Study, 918
Protein polymorphisms, 925–928

R

REES A, BISHOP A & MORGAN
R: Apo(a) gene: structure/function
relationships and the possible link
with thrombotic atheromatous
disease, 873–890
Renal disease, 1014–1016
Resolved and unresolved problems,
865–872
RFLP = restriction fragment length
polymorphisms, 919–920, 922, 935,
943–946, 948–949
RIFKIND B M: Policies for the
prevention of coronary heart disease
through cholesterol-lowering,
1059–1074
Risk factors, 868, 941–959

S

Secondary hyperlipidaemia:
DURRINGTON P N, 1005–1024
Seven Countries Study, 1080
SIDOLI A, GALLIANI S &
BARALLE F E: DNA based
diagnostic tests: recombinant DNA
and cardiovascular disease risk
factors, 941–959
SMITH W C S see TUNSTALL-
PEDOE H
SOUTAR A K & KNIGHT B L:
Structure and regulation of the

LDL-receptor and its gene,
891–916

T

THOMPSON G R: Primary
hyperlipidaemia, 986–1004
Thrombosis & lipid moieties, 871
Thyroid disease, 1012
Treatment of hyperlipidaemia:
ILLINGWORTH D R, 1025–1058
TUNSTALL-PEDOE H & SMITH
W C S: Cholesterol as a risk factor
in coronary heart disease,
1075–1086

U

United Kingdom Study, 929–930,
933–934
United States Study, 931, 933

V

Venn model, 918–919
VLDL = very low density lipoproteins

W

West Germany Study, 931–932
WHO classification of lipoprotein
phenotypes, 986–987
WOOLF N: Pathology of
atherosclerosis, 960–985